The Modular Backyard Power Plant

Ron Melchiore

© 2023 Global Brother SRL

All rights reserved. No part of this book may be reproduced or used in any manner without the prior written permission of the copyright owner, except for the use of brief quotations in a book review.

Disclaimer

This book is designed only to provide information about a modular power plant.

The Modular Backyard Power Plant is sold with the knowledge that the publisher, editor, and authors do not offer any legal or other professional advice regarding this or any other subject.

In the case of a need for any such expertise, you should always consult the appropriate professional, whether that is an electrician, plumber, or other.

This book does not contain all the information available on the subject of modular power plants. It has not been created to be specific to any individual's or organization's situation or needs.

While the authors, editor, and publisher have made every effort to make the knowledge inside this book as accurate as possible, there may still exist typographical and/or content errors that have made it through. Therefore, this book should serve only as a general guide and not as the ultimate source of information on the subject of power plants.

The authors, editor, and publisher shall have no liability or responsibility to any person or entity regarding any losses or damage incurred, or alleged to have incurred, directly or indirectly, by the information contained in this book. The instructions provided have not been reviewed, tested, or approved by any official testing body or government agency; you agree that by using them you waive any right of litigation.

The authors and editor of this book make no warranty of any kind, expressed or implied, regarding the safety of the final products or of the methods used.

Building and operating any of the projects described in this book will be done at your own risk. This includes but is certainly not limited to any physical injury which may occur to yourself or others by incorrectly or unsafely using the knowledge inside this book.

By reading past this point you hereby agree to be bound by this disclaimer, or you may return this book within the guarantee time period for a full refund.

Table of Contents

Chapter 1 - Introduction — 4
Chapter 2 - What's the Big Deal with Solar? — 6
Chapter 3 - Basic Electricity — 8
Chapter 4 - System Components — 11
  Batteries — 11
    Lead Acid Batteries — 11
    Lithium Batteries — 14
    Connecting Batteries — 14
  Solar Panels — 16
  Charge Controllers — 18
  Inverter — 19
  Miscellaneous Components — 20
Chapter 5 - Modular Backyard Power Plant Design Theory — 21
  3-Day Blackout Power Plant Module — 22
  One-Week Blackout Power Plant Module — 28
  The Part-Time Power Plant Module — 29
Chapter 6 - Basic Electric Measurement and Wiring — 32
  Basic Wiring — 32
  Multimeters — 33
Chapter 7 - Let's Build Our System. 3-Day Blackout Power Plant Assembly — 36
  Mounting All Our Components — 39
  Wiring Our System — 44
  Wiring the Power/USB Panel — 47
  Wiring a Light Fixture with Bulb — 50
  Wiring the Solar Panels — 53
Chapter 8 - System Start Up — 56
Chapter 9 - Testing Your System — 61
Chapter 10 - Troubleshooting — 64
Chapter 11 - Helpful Tidbits — 65
Chapter 12 - Materials List and Component Sources — 69

CHAPTER 1
INTRODUCTION

Maine House and Solar Panels

I thank you for your purchase and support of *The Modular Backyard Power Plant*. Let me take a minute to tell you more about myself and what I hope to accomplish.

My wife Johanna and I have been off grid for over 43 years. We've never had a power bill! In those 43 years, most of our adult life, a wire from the power grid has never been connected to a home in which we've lived. All three of our homesteads have been far removed from the power lines. And in that time, we've learned a lot about living a more self-reliant existence with energy production being a big part of that equation.

Late 1970's I purchased my first plot of land in northern Maine and created my first homestead. I was part of the back to the land generation, someone looking for a simpler life. Not long after, Johanna and I became a team devoted to homesteading.

In our first 20 years, we were chronically underpowered. Although I was trained as an electronics technician, understood electricity and the power needs of devices, I didn't have that practical knowledge of matching solar panel size and batteries to what I ultimately wanted to run throughout the day.

Plus, we were chronically underfunded. Money was tight and we simply didn't have the funds to expend on a properly sized system that matched our energy needs. In hindsight, that really should have been a priority since our inadequate system was a source of frustration. Seeing lights go out because we had too many cloudy days coupled with too little battery reserves was exasperating.

Not until I purchased the proper components, hooked them up and used them for a period of time at my locality, did I really understand what it took to build a real power system. At my locality is important. Location is a paramount consideration in how many panels and batteries are needed to power any home on a daily basis.

After 20 years of honing skills along with an on-the-job education, we transitioned to the wilderness of northern Saskatchewan, Canada. We changed locality in a big way. That move gave a new dimension to the term off the grid. We were literally 100 miles in the wilderness on the shoreline of a remote lake accessible only by float plane. We didn't see another human for 6 months at a time and we shopped and picked up mail only twice a year. What a thrill and privilege that experience was!

But we weren't finished with our life's adventures. After 17 years in the bush, we made the move to the coast of Nova Scotia where we have built our final homestead.

All three homesteads were building blocks that gave us an opportunity to improve on what we had. I/we built all

Our Wilderness Homestead

THE MODULAR BACKYARD POWER PLANT

three homesteads from the ground up. I designed and installed our power systems, with each homestead and associated power system being a learning experience while at the same time being a large advance forward. Looking back, I marvel at what a stark contrast there was between my first off-grid system and what we have now. Back then, when I hooked up my first battery and panel, I was jumping up and down with glee if I got 2 amps output on a full sunny day. I quickly found out that was woefully inadequate to power even a small radio and tiny TV. When we installed a few more panels and two marine deep cycle batteries, we tripled our solar panel output to 6 amps on the best day. One would think that by tripling power output, the problem would be solved. But it's not that easy.

Maine is a battleground between seasons and as a result, we had long stretches of cloudy weather in both spring and fall. The battery bank wasn't big enough to give us the power storage needed to get through it. I'm guessing my first car battery was 50 pounds of weight. When I traded that battery in and bought the 2 marine batteries, I might have had 100 pounds of lead acid batteries. Keep that number in mind. When we made the big move to the wilderness of Canada, I vowed we would end our power suffering by installing a much larger system. I put an 800-watt solar array on the roof which on the best day, gave us 24 amps of current, a quadrupling of our Maine output.

And we flew in to the bush almost a ton, a little less than 2000 pounds of batteries. We modernized our home with more lights, satellite TV and more gadgets. We figured our system would easily power everything we had with some leftover. Nope, didn't quite work out that way.

One can make all kinds of calculations, look at solar insolation tables and study climate data. All of that will be a good starting point, but until the house is set up and utilized, one can't be certain everything will work as intended. There were periods during the year when we had all kinds of excess power. But winters in northern Canada are tough and at those latitudes, the days are short. Generator backup was used to get through prolonged cloudy periods.

Fast forward to today in Nova Scotia. We have all the appliances and gadgets of a modern house and are still getting used to this new system. We have a new battery bank that weighs almost 2000 pounds but now our solar array is capable of putting out 100 amps. We've quadrupled our output again compared to our wilderness home. And we have 50 times the current I started out with in Maine when 2 amps was my output. Now we're talking. Progress! We've come a long way from the days of kerosene lanterns and a tiny black and white TV. I'd say that's quite the advance through the years!

To date, we've incorporated all our knowledge into a few different books and we're happy to share and pass on what we've learned.

My intent in writing this particular book and filming the installation process is to provide you with technical knowledge and a measure of confidence so you too can wire up and build your own small power plant to get through any grid down situations. I will explain in simple terms how anybody can understand basic solar electric systems and components.

It makes no difference to me if you are a man or woman, technically proficient or someone frightened at the prospect of simply holding a wire. Together we will go step by step to build a small power system.

Even if you went no further than to read this book and view the video, you will have a treasure trove of information to refer to and who knows, someday you may find yourself building a power system for yourself or a friend. Because this is a portable system, it wouldn't take long to move the system to a friend or family member's home should the need arise. Our modular backyard power plant has a great deal of versatility.

Although my primary focus is to create an energy solution you will use to keep the necessities running indefinitely during a power outage, I would be comfortable using this setup as a solar power source for an RV too.

So, let's define what our objectives are with this book and associated video. By the end of the book and associated video you should know what the individual pieces of a system do, how they relate to each other and ultimately, how you can safely wire the pieces together to form your own basic system.

You should have a better understanding of what any device consumes in terms of power and how to evaluate what energy you will need to produce to power those devices.

Make no mistake about it. This will be a basic system to get you started. Your home will not be lit up like a casino

Off Grid in Saskatchewan

during a power outage. We need to be practical about what you will be able to power with this system and we will talk further about this in chapter 5. It's important to me that you set realistic expectations. Our modular system is a small solar electric power plant designed with expansion in mind. You will be disappointed if you think this system is the equivalent of a small nuclear power plant ready to run the neighborhood.

We will keep this simple to start with. We'll wire up a basic system. We'll try it out and get some real-world experience. The time for testing is now, not when you are in trouble. The basic system will be a tool to learn and evaluate what power devices and household appliances need to run for the durations you'd like to have them run.

We'll run it and if it looks like more power is needed, we can add an extra module. That's the beauty of this system. It is expandable based on your own needs. The parts I selected will allow you to expand in the future as your income and energy needs increase.

This will also be a time to evaluate whether one really needs all the gadgets of a typical household. This will be a help when choosing what priority devices you will need to keep running while the grid is down.

Let's talk about the gadgets and gizmos I envision powering when the utility fails. Certainly, any related to food preservation is a priority. There's too much cost and value in all the food we each have safely stored in our refrigerator and freezer.

As long as we are judicious in opening the doors to these appliances, they will act as cold storage long after the power goes off. But after a day or so, they will need to be powered in some manner.

Other priorities would be any medical devices, a water pump followed by communications such as a portable phone or laptop and finally a few lights.

Don't even think about running an electric cook stove, furnace or the dryer. For one thing, they consume enormous amounts of energy. And for another, they are likely hardwired into the home's electrical circuit breaker box so you wouldn't have access to a plug anyway. On top of that, those appliances may be 220VAC. We must prioritize the most important items.

And finally, I need to mention safety. I am not an engineer or licensed electrician. This work is based on over 43 years of me living off grid with experience designing and wiring multiple power systems for our three homesteads. You are ultimately responsible for building the system and implementing safe practices. We will after all be working with electricity which is safe as long as we are smart about it. I will harp on safety throughout this book and video. Safety is paramount.

You may have heard about alternative energy solutions such as solar electric and wondered what all the hoopla's about. That's a good question! Let's jump right in.

CHAPTER 2
What's the Big Deal with Solar?

Why would anybody wish to build their own solar electric system?

Although I tout, we've never paid a power bill from a utility company, in full disclosure, the upfront cost to produce power on a home scale is significant. Nothing is "free." But once we have a system installed, we can sleep easy knowing our homestead's energy needs are taken care of.

There are many reasons why Johanna and I have chosen to live off grid. We have no worry that the utility will raise our rates. We also have an ultra-reliable source of power with no worries of outages from downed wires or deliberate damage/attacks to the grid. And although our batteries may get low and clouds may prevent charging, the electronics and solar panels that create the power have never failed us in all these years. All of it just keeps working. We have no fear of ever being disconnected or our power being rationed. And best of all, for all three of our homesteads, we had the freedom to choose where we wanted to live, even if it was 100 miles in the wilderness where access to the grid was impossible. How wonderful is that!

Although you live connected to the grid, some of the above reasoning would apply to your situation as well. With your construction of the Modular Backyard Power Plant, you will have a source of power that's ultra-reliable. You will have peace of mind that in the event the utility company cuts power or a disaster takes down the grid for a spell, you will have the power to keep the essentials going.

We live in interesting times characterized by major weather, geologic and geopolitical events. Being able to generate your own power when a disaster strikes gives you a huge advantage over others who are unable to do so.

We can't know whether a hurricane or tornado is around the corner. Will the earth's plates start violently shifting causing an earthquake with concurrent damage to power infrastructure? Forest fires wreak havoc with the power lines as do nuts bent on creating chaos by taking out a transformer or substation. Hydroelectric plants run short of power due to drought and brown outs occur due to seasonal heavy demand. Natural and man-made events will be a continuing theme throughout life which will threaten the many things we take for granted such as water, heat and electric. These things are a year-round worry for people.

We live in Nova Scotia and I started writing this chapter a week after hurricane Fiona hit us. My understanding is this was the lowest barometric pressure system to ever hit Canada. It was a big storm. Florida was hit with a hurricane a week later. Ironically, several months previous to these disasters, my publisher asked me to write a book to help on the grid people deal with power outages.

Once the lights go out, people are left wondering how long before power is restored. It's a crap shoot. Will it be hours? Days? Weeks? For Nova Scotia, the majority of electric customers lost power. A week after the storm hit, there were still many people without power throughout the Atlantic Provinces including Nova Scotia.

No wonder with thousands of trees down across power lines and poles snapped off. It was a massive effort to restore power to all those customers. You might ask aren't most people prepared for temporary outages with a back-up generator in place? The plan being to just run the generator to get through until power is restored. Sure, that's a great solution for a local outage of short duration.

But, when a large swath of the Province was devastated and had no power, it became a game changer. Gas stations who didn't have backup generator were temporarily out of business and useless for securing gas. The gas was in a tank below ground with no means to pump it to needy customers. Thousands and thousands of generators across the Province were all running. The service stations that either never lost power or had backup generators had car and truck lines stretching down the road with vehicles desperate for fuel not only for their cars and trucks but for their generators as well. Our local gas station ran out of gas. The nearest major town an hour away from us with at least a half dozen gas stations all ran out.

Anybody who had a modest reserve of gas never anticipating days of lost power was now scrambling to get gas to keep the generator running. The refrigerated/frozen food supply was on the line if that generator didn't keep purring until power was restored. A lot of people were hitting the panic button. People had to drive hours to find a gas station that actually had fuel. Quite the gamble since the car would now need to be refueled as well and if they were unable to buy gas, they were up the creek.

As a society, we sure are dependent on electric power. We are also dependent on being able to drive to the local gas station to grab a can of gas. When both become unavailable, that's a big problem. Who wants to have to worry about that? Certainly not I. To eliminate this worry we will build a simple, affordable solar power plant that is expandable depending on each home's needs. Together we will remove the need to be dependent on outside forces to get through a power outage regardless of its cause.

Johanna and I are not preppers but as homesteaders who once lived for 17 years on a remote lake far in the wilderness, we learned to be as prepared as anybody can be. We were some of the early adapters of off grid technology so you can have some trust that I know what I'm doing.

When I was asked to write a book to help folks like yourselves bridge the gap between a power outage and restored power, I put myself in your place. I'd want the best bang for my hard-earned money. If from experience using the basic starter system I found it didn't supply all my energy needs, I'd want a simple way to add more batteries and solar panels as additional modules to increase my power output. And I'd insist on reliability. If the system lets me down when I need it, it would be useless to me.

By nature, I'm a thorough guy. Whether it's building a new homestead, planning a garden or trying to solve a problem, I take time to understand the problem and then I take all the time needed to research, think, research some more and ponder until I am convinced I have the best game plan for success.

For weeks I researched components trying to come up with a system I would want for our own use if we were hooked up to the grid and concerned about any sort of power disruption. I sent out emails to many different solar supply companies with generic text inquiring about solar power components. That was my way of testing these companies to see what kind of response I got. If they were prompt and thorough with their answers, they passed the test. For those who responded, I then leveled with them and told them what I was really up to. Obviously if a company couldn't be bothered to respond, we don't want anything to do with them.

Out of all the companies I wrote to, three stood out. The rest didn't bother to respond. I'll supply you with a materials parts list as well as sources to start with for your shopping at the very end of this book.

So, the question is: When the power goes out, what system can you deploy to keep the lights on, the refrigerator/freezer cranking and recharge cell phones and flashlights? The Modular Backyard Power Plant!

I'm going to be with you step by step via written word as well as video which will be a visual aid to those who find seeing things facilitates comprehension. Together, we will wire the entire basic system. Then we will add a module to double the capacity. If that's not enough, we will add a final third additional module to double our energy capacity again. Bottom line, by the time you get to the end of this book and video, if you have all the components specified and have them wired properly, you will have a functioning off grid power system. If the power goes out - no problem. Pull the Modular Backyard Power Plant out, set it up and plug those appliances in.

So, let's get started but first take a deep breath. Everything's going to be OK! You're in good hands.

Oh, and in case you were wondering, we never lost power during hurricane Fiona. Our off-grid solar electric system kept running as if nothing had ever happened. Our frozen and refrigerated foods were fine, the lights were shining and we had water and communications without any interruptions. The reserve of gasoline we always have on hand was at the ready to help neighbors should they run short before being able to get fuel for their generators.

CHAPTER 3

Basic Electricity

Before we get too involved, we should really take some time to discuss some basic electrical terms and practical theory. You don't have to master this stuff completely but it will prove invaluable to you and give you a better understanding of what goes into system design. All I'd like you to do here is grasp the concepts. Then use this chapter as reference when it's needed.

The following are terms we will be using along with its symbol and measurement

Voltage (E) measured in Volts (V)
Current (I) measured in Amps (A)
Resistance (R) measured in ohms (Ω)
Power (P) measured in Watts (W)
Kilo (k) = 1000

You may need to measure a voltage or check for continuity and knowing a few things ahead of time should be a big help. And at dinner parties, it's always nice to drop a line like "yeah, I just measured my battery bank voltage with my digital multimeter… it's 24.72 volts- a little low but sun should bring that up tomorrow." You'll be the talk of the party!

The two units of measure we will deal with now are voltage (E) in Volts (V) and current (I) in Amps. We've all heard the terms but let's face it, most don't have any idea of what they mean. The easiest way to think about this is to make an analogy with a water pump connected to a pipe. We are all familiar with a water pump. It's a device that moves water. If we are going to pump water, we really want the ability to direct that water to a place where it can be used. That's through pipes. Think of all the different pipes in a house supplying water to utility sink, shower, toilet, clothes washer, kitchen sink etc.

Now let's make the mental conversion to a simple electric circuit. Instead of a water pump, we have a battery. The battery becomes the water pump. And much like a pump creates pressure to push water down a pipe, a battery provides the force to push current down a wire. Think of wires as pipes. With me so far?

I'm sure you know that water pumps come in all sizes. A pump might be rated at 1 gpm and another might be rated at 100 gpm. GPM is a measure of how much water can be moved by the pump and it stands for gallons per minute. Obviously the bigger the pump the more gallons of water it can force down a pipe in a minute.

That is true of batteries as well. They too come in all sizes and voltages. The bigger the water pump, the more water it can push, the higher the voltage of the battery, the more current it can push down the wire.

By now, you've also thought about the different sized pipes you've seen around. Your home is a good example of having many different sized pipes. It stands to reason, the bigger the diameter of the pipe, the more liquid it can carry. When you go to your favorite fast-food shop for a soda, it comes with a straw. That's a good diameter to suck up liquid from the cup. But if there was a house fire, you wouldn't want the fire department showing up with fire hose the diameter of a straw. For one thing, a fire needs as much water as possible being poured on it to contain it, and second, if you connect a big pump to a straw, that straw is going to self-destruct from the force created by that big pump.

A wire is the same concept. Wires too come in different diameters. I've included a standard wire chart here. We will reference back to it a few times in our discussions. You'll note that the higher the wire gauge (size) number, the smaller the wire diameter. Opposite what you would think. The bigger the diameter of the wire, the more current can be pushed through the wire, just as a bigger pipe allows more water to flow. And just like the straw example, if you try to push too much current through a tiny wire, it will simply heat up and then burn up.

In our modular solar electric kit you will note wires of different thickness. Wire thickness is an extremely important consideration when designing or wiring a project together. The capacity of the wire and how much current it can handle must be factored in in order for everything to be safe. I've done the thinking for you here and the wires are the right diameter for our power plant.

The final piece of this analogy is to tie in the terms voltage (E) measured in volts (V) and current (I) measured in amps (A). A water pump creates pressure to force water down the pipe. In our electrical system, voltage (E) is the equivalent of pressure and current (I) is the equivalent of water flow. Take note that I am incorporating the correct symbols in parentheses after some of these terms. We will use the symbols as shorthand further on.

Most have heard the term power (P) and watts (W). Remember the old fashioned incandescent light bulbs rated in watts? A 60-watt bulb or perhaps a 100-watt bulb were common in house light fixtures. The watt rating was how much energy the bulb was consuming while being powered. How does one determine watts?

Very easy! Multiply voltage (E) times current (I). If you multiply voltage X current you will get power (P) in Watts (W). The formula is P= EI So, for example, let's say you have a 12-V battery connected to a light bulb. If the 12-volt battery is forcing 1 amp to flow through the wire to make the bulb light up you can determine the power in watts using the formula P=EI as follows: 12V X 1A = 12 watts. So the light bulb is consuming 12 watts of power. In this example, there are two more pieces to the puzzle we can add.

American Wire Gauge Conductor Size Table

American wire gauge (AWG) is a standardized wire gauge system for the diameters of round, solid, nonferrous, electrically conducting wire. The larger the AWG number or wire guage, the smaller the physical size of the wire. The smallest AWG size is 40 and the largest is 0000 (4/0). AWG general rules of thumb - for every 6 gauge decrease, the wire diameter doubles and for every 3 gauge decrease, the cross sectional area doubles. **Note** - W&M Wire Gauge, US Steel Wire Gauge and Music Wire Gauge are different systems.

American Wire Gauge (AWG) Sizes and Properties Chart / Table

Table 1 lists the AWG sizes for electrical cables / conductors. In addition to wire size, the table provides values load (current) carrying capacity, resistance and skin effects. The resistances and skin depth noted are for copper conductors. A detailed description of each conductor property is described below Table 1.

AWG	Diameter [inches]	Diameter [mm]	Area [mm^2]	Resistance [Ohms / 1000 ft]	Resistance [Ohms / km]	Max Current [Amperes]	Max Frequency for 100% skin depth
0000 (4/0)	0.46	11.684	107	0.049	0.16072	302	125 Hz
000 (3/0)	0.4096	10.40384	85	0.0618	0.202704	239	160 Hz
00 (2/0)	0.3648	9.26592	67.4	0.0779	0.255512	190	200 Hz
0 (1/0)	0.3249	8.25246	53.5	0.0983	0.322424	150	250 Hz
1	0.2893	7.34822	42.4	0.1239	0.406392	119	325 Hz
2	0.2576	6.54304	33.6	0.1563	0.512664	94	410 Hz
3	0.2294	5.82676	26.7	0.197	0.64616	75	500 Hz
4	0.2043	5.18922	21.2	0.2485	0.81508	60	650 Hz
5	0.1819	4.62026	16.8	0.3133	1.027624	47	810 Hz
6	0.162	4.1148	13.3	0.3951	1.295928	37	1100 Hz
7	0.1443	3.66522	10.5	0.4982	1.634096	30	1300 Hz
8	0.1285	3.2639	8.37	0.6282	2.060496	24	1650 Hz
9	0.1144	2.90576	6.63	0.7921	2.598088	19	2050 Hz
10	0.1019	2.58826	5.26	0.9989	3.276392	15	2600 Hz
11	0.0907	2.30378	4.17	1.26	4.1328	12	3200 Hz
12	0.0808	2.05232	3.31	1.588	5.20864	9.3	4150 Hz
13	0.072	1.8288	2.62	2.003	6.56984	7.4	5300 Hz
14	0.0641	1.62814	2.08	2.525	8.282	5.9	6700 Hz
15	0.0571	1.45034	1.65	3.184	10.44352	4.7	8250 Hz
16	0.0508	1.29032	1.31	4.016	13.17248	3.7	11 k Hz
17	0.0453	1.15062	1.04	5.064	16.60992	2.9	13 k Hz
18	0.0403	1.02362	0.823	6.385	20.9428	2.3	17 kHz

Wire Gauge Chart Courtesy of Solaris

Using the 12-watt light bulb as an example above, let's assume that the light bulb was on for exactly 1 hour. We can now say that the light bulb consumed 1 amp-hour AH (it drew 1 amp for 1 hour) or 12 watt-hours WH (it drew 12 watts for 1 hour) of energy. Watts is a measure of power drawn with watt hours a measure of power drawn over time, also known as energy.

Although this seems like a lot of mumbo jumbo, it's important to grasp at least a rudimentary understanding of these terms and how they relate to each other. And here's why. You will be wiring a solar electric system together. It is a finite system meaning you can't just plug the rest of the world into it and expect it to power everything. There are limitations to what any electrical system can do. Even the grid which we think of as infinite really isn't. Ask anybody who has experienced a brown out or black out due to high electrical demand. This will make more sense as we go along.

Sometimes voltages and power are large values. That's where the kilo (k) comes in. Your utility power bill is probably measured in kilowatts and you probably pay based on how many kilowatt hours you have used in the month. Kilo just makes it a little easier than writing out large numbers. Instead of 10,000 watts, you can also write it out as 10kW (10 kilowatts). Instead of 100,000 watts, it is also 100kW (100 kilowatts).

We've covered a lot of what may be confusing, unfamiliar stuff. That's OK. Sleep on it. Re-read it again and it will start to make more sense.

Let's take a recess. Go for a short walk and grab a drink (not that kind of a drink! I need all your brain cells focused on this).

Okay, now that we've rested the brain cells, we're ready for more mental stimulation. We've gone over some basic electrical terms. One of those terms is voltage. I'm guessing you've all heard that voltage comes in two distinct forms. AC (alternating current) and DC (direct current). There's no need to bog down on this. But you definitely need to know the difference. Alternating current (AC) is grid power. Just about all house appliances are made to run on 120VAC which is the power that runs in from the electric company. Your home appliances plug into an AC wall outlet. A generator is another source for AC power. The generator is made to produce an alternating current so that any AC appliance cannot tell whether it is running from the grid or from the generator. The voltage is the same. In North America which is the primary audience for this book, the standard AC voltage is 120 VAC at 60 hertz.

If you had a way to visually look at alternating current (AC), it would be waves and valleys, peaks and troughs. Hence the name alternating. The frequency or how fast that alternating change takes place is measured in hertz. Hence the North American grid has a frequency that alternates at 60 hertz, 60 times a second.

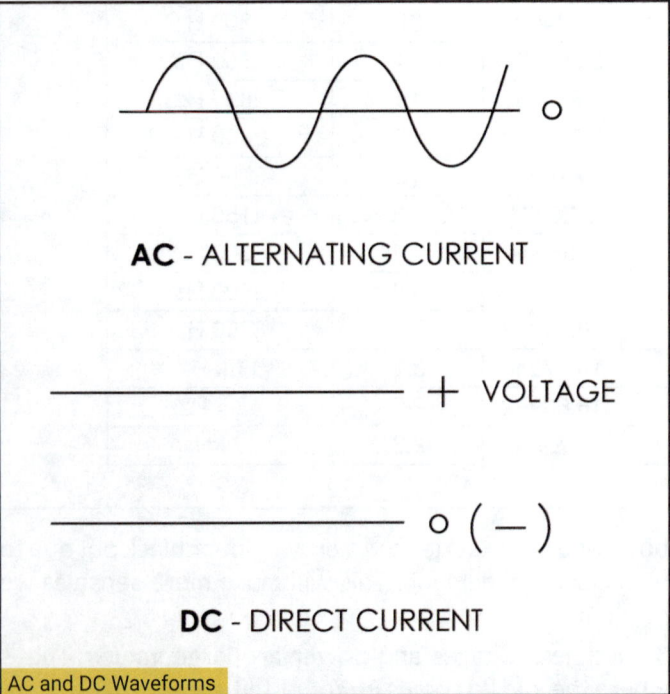
AC and DC Waveforms

Then there's DC voltage which is direct current. This is the current provided by batteries or solar panels. If you could visually look at a DC voltage, it would be a straight line, there would be no waves because there is no frequency.

At this point, all I want you to know is we will be working with both AC and DC. A device or appliance made to run on AC must only be used with AC. The same applies to DC. You can't wire an outlet with DC battery voltage and then plug your AC refrigerator in and expect it to run.

Don't even try something like that unless you want to see that nice refrigerator go into self-destruct mode. In the next chapter, we will talk about a device that will take DC voltage and convert it to AC so that you can run your refrigerator. That device is called an inverter.

I haven't mentioned anything about resistance. This is the last major term we'll discuss in this chapter. Stick with me, you're almost through with the technical material. Resistance in reference to electricity, quantifies how a material reduces the current flow through it and is measured in Ohm's. Voltage (E), Current (I) and Resistance (R) are all related to each other. Ohm's Law are mathematical formulas showing the relation to each other.

Voltage	Current	Resistance
V = I.R	I = V/R	R = V/I

Ohm's Law Triangle

Going back to our water pipe analogy, the bigger the water pipe, the less resistance and the more flow. That makes sense. If you want to move a lot of water, get a bigger pipe. But let's look at it a different way. Let's start making that hose narrower. We're making it more "restrictive". We're creating more resistance and making it harder for the water to flow out of the pipe. And the longer the hose, the more resistance.

If you had a short piece of ½ inch diameter garden hose and turn the faucet on, the water will come out with force. Now connect 200 feet of the same ½ inch diameter garden hose. The water at the end of the line isn't coming out with as much force. If you had a mile of hose, it might be reduced to a mere trickle. What changed? The diameter of the hose stayed the same, it's only the length of the hose that changed.

The same is true for electricity, the bigger the wire the less resistance. The bigger the wire, the cross-section, the easier it is for current to flow and the more current it can handle. When you put 12.0 volts at one end of the wire, you ideally want to see 12.0 volts at the other end of the wire. You don't want to see the "voltage drop" to perhaps 11.2 V You don't want to lose voltage along the way.

You can see how important it is to use the right wire size or "gauge". When selecting the right gauge of wire, it's a balance between the cost of the wire, ultimate safety and how much line loss (voltage drop) you are willing to accept. Line loss is just as the name implies. How much voltage are you losing down the line and what is the voltage that ultimately shows at the end of the line.

Just a tid bit for you. When you lose voltage at the end of the line, it seems to disappear. But what's really happening is the voltage that's being forced down the line is fighting through any resistance that the line has. And that fight creates friction which generates heat. That's why using too small a diameter wire for the current being pushed through it might potentially cause the wire to burn up.

It's nice to talk about voltage, current and resistance in theory but there may be times when you need to actually measure a voltage, current or resistance. A multimeter is a tool that can actually provide measurements of volts, amps and ohms. It is a very handy device to have around the homestead. I can't imagine not having my multimeter with me. You can buy a cheap analog meter with a moving needle or a cheap digital meter. I'd opt for the digital meter because it will be easier to interpret the numbers. We'll delve further in how to use the meter after we've built our system.

You've made it through. Good job! I tried to make this as pain free as possible. The good news is there's no final exam!

CHAPTER 4: System Components

There are four main components to our solar installation and many minor parts. The four main components are: batteries, solar panels, charge controller and inverter.

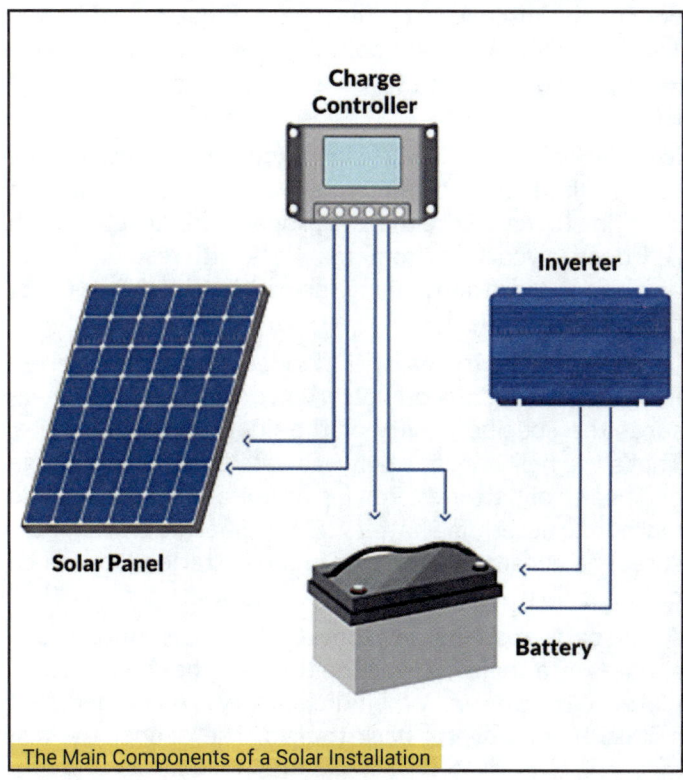

The Main Components of a Solar Installation

We'll break these down one at a time and discuss them in detail.

Some of the minor components but certainly not all inclusive are cables and wires, electrical boxes, metering devices, fuses and circuit breakers. I don't want to minimize the importance of the minor parts because most are absolutely necessary for the operation of our power plant.

Batteries

Batteries are the energy storage part of the system and come in many different voltages. The most common battery voltages used in off grid systems are 2V, 6V, 12V, 24V and 48V. The amount of energy that is stored in a battery is typically rated in amp hours (AH) or watt hours (WH).

Batteries contain a mixture of chemicals, otherwise known as battery chemistry, which can be converted to electrical energy. And as the years go by, that mixture of chemicals is evolving to become more efficient and better for use in different applications due to the advancement in battery chemistry.

The batteries are the means to store energy for later use. When you turn the key to start a car, you rely on a battery which has enough energy stored to do the job of turning the engine over. Once the job of starting the car is done, an alternator is in place to recharge that battery. The same concept applies to our solar electric plant. When the power grid goes down, we will deploy a system to run our necessities with the capability to recharge it.

Lead Acid Batteries

In Maine, I made the mistake of buying a car battery as my first energy storage. It was a flooded lead/acid type. (typical car battery) Back over 40 years ago, there weren't many choices for battery types. While a car battery will work, it doesn't work well. The problem is with the design of a car battery. They are made to give a quick burst of current to turn the engine over for starting. They are not made to be deeply discharged and then recharged. Doing so shortens their life expectancy dramatically.

What we want for our backyard power plant is a battery that is made for deep cycling. After I killed our car battery in Maine, I bought marine deep cycle batteries. Although marine deep cycle batteries are a step up, they still are a poor choice for off-grid. What you want is a deep cycle battery made to take the punishment of being drained deeply and then recharged. True deep cycle flooded lead acid batteries fit the bill.

Along with the flooded lead acid batteries, there are AGM (Absorbed Glass Mat) and Gel batteries. Those versions are sealed lead acid batteries with different life cycle characteristics and pricing as compared to the flooded batteries. They are more expensive than flooded, but the big advantage they have compared to the flooded cell is that they are maintenance free. Flooded cells need to be checked and over time, water needs to be added whereas the AGM and Gel do not. But I don't find this a big deal really.

While I'm on the topic of lead acid, I'll continue with my experience with them since I've used them for 43 years. If you understand a few things about the basic lead acid battery first it will make any comparison with newer battery types easier.

I've had wonderful results once I committed to a proper deep cycle battery capable of thousands of cycles. In our wilderness homestead in Saskatchewan, we flew in almost a ton of lead acid batteries. When we departed after 17 years, those batteries were still going strong. I am confident the batteries we had in the bush would have lasted 20 years or more had we stayed there.

We moved to our current location in Nova Scotia about 6 years ago and based on costs and past experience, we again chose large lead acid 2 Volt cells. We have a 2000AH battery bank that cost roughly $7000USD. I expect 20 years of life at a minimum. If we get the life out of these batteries I am expecting, that works out to a cost of $350/year for our batteries. (7000/20=350) That seems pretty cheap to me.

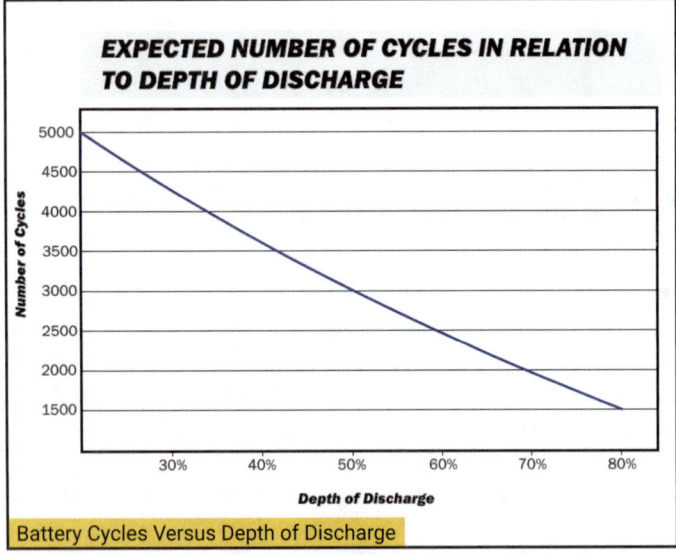
Battery Cycles Versus Depth of Discharge

Or another way we can evaluate this is as follows. According to our battery life cycle curve shown below, even if I was to discharge our batteries down to 50% every cycle, we should still expect 3000 cycles. Our battery is 2000AH@24V which equals 48,000Wh (2000 X 24= 48,000).

Our purchase price was $7000. We expect 3000 cycles at a minimum. That works out to $2.33/cycle. (7000/3000 = 2.33) If my expectations are met that I will far exceed 3000 cycles, the numbers are even further into the win column in my favor. The cost would be even cheaper for us.

I know a lot of battery connoisseurs will object to the conclusion that lead acid still have a place in an off-grid setup.

They will rightfully list all the benefits of the other battery technologies such as lithium phosphate or lead carbon. And they would be right. My only argument is cost of the new technologies versus life expectancy. I have a problem with the models used to project cost versus life benefit of the other battery types. Assumptions are made that the lead acid battery will be drained the same depth as a competing battery technology. If you make that comparison, the lead acid battery loses.

As long as one understands how to get more life out of the lead acid battery, I would argue, the lead acid still has merit in an off-grid situation.

How do you get more life out of a lead acid battery? The secret is to shallow discharge. I try not to take my batteries below 20% discharge before charging them back up. If you refer back to the previous depth of discharge curve versus lifetime cycles chart, you will see how the deeper the discharge the less life your batteries have. The curve for our batteries shows we can expect 5000 cycles at a 20% discharge.

Five thousand cycles is almost 14 years. That's of course if the information from the battery manufacturer is accurate. I would hope a reputable company is in the ball park with their data. Add in the fact that in spring, summer and fall there are many consecutive sunny days where our batteries might only discharge 5%. Inside of a couple of full sun hours, the batteries are fully charged. That has to add years to battery life.

When selecting a battery, ask for the graph that shows life expectancy versus depth of discharge. This will tell you how many cycles the battery can handle before it needs to be scrapped. A cycle refers to any time the battery is discharged and then recharged. The more cycles and the deeper the discharge, the less life the battery will have. This is a really important point to understand. If you shallow cycle your batteries, you will get far more life out of them.

In addition to shallow cycling, another secret to long battery life is the maintenance. Maintenance consists of checking to make sure there is proper water levels in each cell, doing a yearly hydrometer check with results written down as a record and making sure the terminals and top of the battery are clean.

If you ever have to deal with flooded lead acid batteries, I suggest buying a good battery hydrometer which measures the specific gravity of the liquid in the batteries. That's the best way to really evaluate the state of charge and health of batteries. Relying on voltage is not a reliable method to determine battery health. Specific gravity measures the density of the liquid in the batteries. How much of the liquid is water and how much is acid.

As the battery discharges, some of the acid is converted to sulfates by a chemical reaction that attaches it to the lead plates of the battery. When the battery is recharged, that sulfate then converts back to acid. The deeper the discharge, the more sulfate forms on the plates. Over time, some of that sulfate remains on the plates and builds up. One of the things I did as soon as I got new batteries is I numbered them with masking tape and a magic marker so that for the next 20 years or more, I'll always know which battery is which.

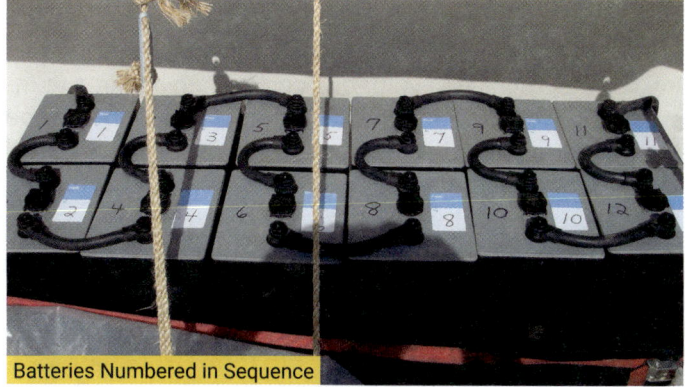
Batteries Numbered in Sequence

I made a sketch of the numbered batteries in a notebook and then recorded the specific gravity of each cell along with the date. No matter what type of battery you have, it's not a bad idea to number and/or date your batteries with a magic marker so that you have a record of them.

In order to receive the best support from any battery manufacturer, it's best if you can prove to them you are a pro with batteries and this is one way to do it. Each year, take specific gravity readings and continue recording these numbers through the years.

You will hear the term equalization in regards to batteries. Equalization is a forced charge into the battery with a higher voltage. That sulfate that collects on the plates during discharge sometimes doesn't all revert back to acid as it should. When you read specific gravity, the readings should all be very close to one another and should be close to the previous readings. If they are not, it's time to equalize and try to bring all the cells back in line with each other. By forcing current at a higher voltage, we are trying to clean the plates of sulfate and bring all cell specific gravities together again.

The downside of lead acid batteries are many which is why the newer technologies are changing the off grid landscape. Obviously, they contain lead which is not only heavy in sheer weight but is also toxic. Perfectly safe while contained in the battery but once the battery is dead, one wants to make sure it is recycled properly.

A lead acid battery contains sulfuric acid. Once again, safe as long as it remains in its container. Lead acid batteries as I've mentioned have a limited life span if they are discharged deeply. Taking them down to a 50% discharge, half empty, is about as far as you want to go with them. Can you take them to stone dead? Of course. Who hasn't had a car battery that was dead. Give it a charge and it's back to working. But doing something like that on a regular basis dramatically shortens the battery's life.

During the charging process, lead acid batteries off-gas. Hydrogen and Oxygen typically. If you can smell the batteries charging, you probably smell hydrogen sulfide which isn't healthy. That will only occur when the battery is reaching full charge or is being overcharged.

Because the charging process is breaking the liquid electrolyte down into gasses, over time, the batteries will occasionally need to be maintained by adding distilled water to keep the plates in the battery covered. I'd highly recommend you do NOT use tap water to top off your batteries. Tap water is not pure and one cannot know what chemicals and minerals are in that tap water. Use only distilled water which has gone through a purifying process when flooded lead acid batteries need more liquid.

In Saskatchewan, we always had our flooded lead acid batteries in the house and I had them located in a ventilated room. Now our batteries are in the basement and in order to comply with code, I needed to house them in a tight box that is vented to the outside. I planned ahead for this when we poured the concrete for our ICF (Insulated Concrete Form) basement wall by putting in a vent pipe. That vent pipe goes to the battery box.

Vented Battery Box

A fan can be purchased specifically for this venting application and is installed in the vent pipe. The fan would be connected to the battery charger's auxiliary output.

The purpose of the fan is to vent any gases the batteries produce while charging. Batteries only gas a lot when they are getting close to fully charged or when they are being equalized. The chargers these days are smart enough to know the relative state of the battery and they will automatically cycle the vent fan on when it is needed to exhaust battery gas to the outside. The AGM and Gel sealed batteries don't have this concern. Because they are maintenance free, they don't produce gas which is an advantage if batteries are located in the house.

While we are on the subject of maintenance, we will want to occasionally check our terminals and battery tops to make sure they are clean. If you've ever seen a car battery with some corrosion on the terminals, you know it causes poor connections. I use a damp to wet rag with baking soda to clean our terminals and battery tops. I'll make a solution of baking soda mixed with water. I have no set recipe and use my judgment depending on how much corrosion I have to deal with. The baking soda neutralizes the acid. It is imperative none of that baking soda ever finds its way into the battery cells. The caps must be tight before doing any cleaning.

Any time I work with batteries, I wear goggles or safety glasses. That will be mentioned more than once in this book. We only have one set of eyes and we will protect them at all costs. Battery electrolyte is a liquid and can and will splash especially if one is not careful resulting in potentially dangerous burns. Taking specific gravity readings is one example of when we are dealing directly with the liquid. Have a source of water handy in case you splash some acid on you. Wash it off immediately with lots of water. Take off rings and any chains before starting to work. It's safer to work around batteries with no metal that could potentially short the terminals.

That gives you a better sense of the pros and cons of the basic lead acid battery. These days, there are new battery compositions and promising new battery technologies. Improvements to existing technologies are constantly being made. Although I can still make the case to use flooded lead acid batteries for off-grid application, honestly, as time goes on, as improvements are made and as prices come down for lithium, the argument for lead acid batteries is rapidly becoming less compelling. Nevertheless, I wanted to share our lead acid battery experience with you as both an education as well as to give you the reasoning behind why I selected the battery I did for the Modular Backyard Power Plant.

Lithium Batteries

In fact, I have chosen lithium phosphate batteries for our power plant. Many of you might wonder why I just didn't go with cheap lead acid batteries for this application. First the lithium phosphate batteries are relatively light meaning they are energy dense; they pack a lot of power in a small package. I want to make sure our project can be built by anybody, not just weight lifters.

While Johanna and I have to monitor and be careful about the depth of discharge with our lead acid, that's not the case with lithium batteries. Instead, you can take them to 80% discharge and not degrade the life of these batteries. And depending on the manufacturer, you can get impressive life cycles out of them. The lithium battery I selected has a 3500-cycle rating at 80% discharge. Or put another way. You can discharge the battery to 80% and then recharge back to full at least 3500 times. Another advantage over the lead acid battery.

I also chose lithium because you will not have to worry about any off gassing, maintenance or equalization charge. All batteries will self- discharge over time but these batteries keep more of their charge each month. In other words, you can store lithium batteries long term and not worry as much about whether they are fully charged when it's time to use them.

However, it's best to be seated when pricing these batteries. They are a significant investment which is why at least 6 years ago when they were even more expensive, we went with flooded lead acid batteries here at our Nova Scotia homestead. Down the road, if you ever find yourself in the situation of going off grid or wishing to have a grid tied system with battery backup, it will be worth doing the research and evaluating costs when it's time to install your system. Things change rapidly and there may be another form of battery that makes more sense.

Connecting Batteries

I'd like you to think of batteries as building blocks. We can hook up many batteries together to form larger batteries. That's what I did here in Nova Scotia and at the Saskatchewan homestead. Instead of one 24-volt battery which would have weighed almost a ton, I bought twelve 2 volt, more manageable cells and connected them in series (12 cells @2V = 24V).

BATTERY OR SOLAR PANEL CONNECTIONS

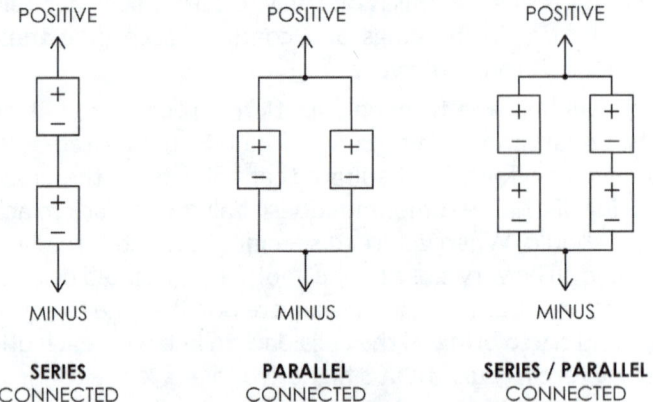

Looking at the battery connection illustration, you can see series batteries are connected with the positive of one battery going to the negative of the next battery in line, in series. When you evaluate the whole package as one battery, you can see that there is still one open positive terminal on one of the batteries and a negative terminal on the second battery. More batteries connected the same way still gives you a positive and negative terminal.

And batteries can be hooked up in parallel too. The positive terminal on one battery can be connected to the positive terminal of another battery. The negative terminals are connected between the two batteries. They are parallel to each other.

Lastly, and this is a mouthful, strings of series wired batteries can be paralleled. We won't be doing that but I wanted you to know it's possible. The one caveat is we will always follow the manufacturer's recommendations. If they say don't wire our batteries in series, we make sure we don't.

If you're stressing over this, don't. In our power plant, you don't have to worry about this stuff. We keep it simple. But it's still good information to know since this book/video will be a valuable resource for those wanting to know about solar electric systems in general.

Batteries have specifications and rules we must work with. Regarding lead acid batteries, the one rule I felt important was to always have my batteries the same age and capacity. That way, if I ever had trouble with a battery, I would know they were all purchased at the same time. And because the batteries were all the same size, they should all discharge and recharge somewhat equally. You can see if I had say a 100AH battery wired with a 200 AH battery, that poor 100AH battery would be getting overcharged while the 200AH would still happily be taking a charge. Not a good situation.

Regardless of battery chemistry, I think this is still a good rule to follow. Keep the age of your batteries as close together as possible and have batteries all the same capacity. However, in researching this, it seems there is quite a range of opinions about this. What it will come down to is following the manufacturers advice.

These lithium batteries come with their own BMS, battery management system, which is a means of monitoring the battery. Basically, a computer is part of the battery. Each manufacturer has their own BMS and it is best to inquire of tech support if you were thinking of doing something non-standard. Again, this not anything we will be doing with our power plant. We have selected batteries that will play nice with each other so that there's no conflicts. But I wanted you to be aware of potential conflicts if you were thinking of mixing different manufacturer's batteries together in a system.

The specifications we care about at this stage are volts and amp hours. The rule is this: when batteries are connected in series, each battery voltage is added up to make the total bank voltage. Current in amps will be constant in a series connection. The amp hour rating of the battery bank will be the same as any one battery in the string.

For batteries connected in parallel, the reverse is true. The voltage stays the same but the amp hour capacity of each battery is added up.

And then there's that mouthful we mentioned a few paragraphs back about batteries hooked in a combination of series/parallel, but there will be no need for a connection like that in our power plant.

The easiest way to explain this is through examples. The battery for our examples is a 6V 100AH battery.

It is imperative to understand that a battery has two terminals. A plus, positive terminal, usually symbolized with a plus sign (+) and a minus, negative terminal, usually symbolized with a (-). In DC circuits, the general wiring standard is to make a red colored wire the positive connection and a black colored wire would be the negative connection. When we make any connection to a battery, I need to stress that the connection must be right when it is done. Before you would ever wire a battery, stop and think it through. Be certain the connections are being made exactly as specified in any wiring diagram. Careful attention must be made to plus and minus when connecting batteries. It would be prudent to wear safety glasses when working around batteries. I don't want you to be fearful, it's just a good safety practice to get into. I am naturally over cautious in my own life and I pass safety on to you as well.

In our series connection, you can see each 6-volt battery that is connected in series, adds 6 volts to the total. So two batteries in series makes a 12-volt battery, four batteries in series makes a 24-volt battery. The current flowing through the series batteries will be constant depending on the load put on the battery. So in other words, if a light bulb draws 1 amp, anywhere in that series connected battery, the current will be 1 amp.

Typically, off-grid battery bank voltages are either 12V, 24V or 48V. Remember the water pipe analogy. The bigger the pipe, the more water can flow. The bigger the wire, the more current can be pushed through it. In a 12 V system, you need bigger wires to supply power to a large load. The higher the battery voltage, the smaller the wire diameter can be for that same sized load.

So, a 12 VDC off-grid system is fine for a very small system and 48V is best for a larger system. I favor 24VDC for the middle of the road.

A battery is capable of producing a certain amount of energy for a given period of time before it becomes fully discharged. For example, let's assume I have a 100AH battery. *Theoretically*, that battery will be able to supply a load of 100 amps for an hour and then it's fully discharged. It can supply 50 amps for 2 hours before being fully discharged. You can see the progression of how this works. A 1-amp load could run for 100 hours.

Note I underlined the word theoretically. A major factor in how this works in real life is how fast you drain the battery. The bigger the load, meaning a lot of energy is being drawn from the battery in a short time span, the less battery life you really have. The lighter the load, the easier it is on the battery and the more you will get the full amount of energy available from the battery.

I'm a masters sprinter now 67 years old. The muscles in my legs have lots of stored energy. You want me to jog for an hour, no problem, I'll run for miles. You want me to sprint 200 meters, again, no problem. But the second or third 200-meter sprint in a short period of time with little or no recovery is taking a big toll and I'm struggling by the third sprint. I'm expending a lot of energy to sprint 200 meters and my legs are getting drained rapidly. If I was to sprint 200 meters at half speed and then get a good rest and then repeat, I could do that many times before my legs are shot. It won't add up to miles but if I run my 200 meters at half speed with a rest in between, I'll ultimately run quite a distance in an hour as opposed to being played out completely by my third 200 meter full out. Same principal applies to the battery.

Now let's assume I've powered my load using my 100AH battery and now my battery is 100% discharged. In order to charge it back up, we need to have a charging source such as a solar array capable of putting energy back into the battery. If you have a solar array that only puts out 1 amp, then it will take 100 hours of full sun to recharge the battery. If your array puts out 50 amps, it will take 2 hours to recharge. One hundred amps and your battery is charged up in 1 hour. Again, this is a theoretical over-simplification.

I don't want to have your eyes glaze over but know that nothing is 100% efficient so in reality in order to recharge the battery, it will actually take more time than what I stated. So, for example, if I drew 100AH (amp-hours) out of the battery, it might take 110AH of solar/wind power to recharge the battery because some of that energy from the sun is electrical energy while some is converted to heat and some is lost to the device tasked with making the conversion from sunlight to battery charge. Keep the concept in mind and we'll discuss it in more detail when appropriate.

Solar Panels

Solar panels convert sunlight to electrical energy. They are the means we will use to recharge our batteries during daylight. Solar panels produce direct current (DC) voltage, a form of energy batteries can accept. There are different technologies with Monocrystalline and Polycrystalline being the common choices for panels. Just like our batteries, technology is constantly improving both versions. Both technologies work equally well these days so what you choose will come down to bang for the buck. What panels are on sale and what is the price per watt. As of this writing, the 200-watt panels I've chosen to use for the power plant are $250 each. That works out to $1.25/watt.

There are plenty of larger panels made that will be quite a bit cheaper than that and they would be fine for a permanent off-grid set up. But for a portable set up like the Modular Backyard Power Plant they aren't the best option. I also had the choice of flexible and foldable panels. I chose our panels because they are rigid with aluminum frame, are more durable, sturdy and easy to transport inside to outside when needed. Portability and ease of moving them is also a consideration if you are concerned about overnight theft of your panels. The wires are a cinch to disconnect and each panel in your system can easily be brought indoors overnight.

I know what some of you clever people are thinking. You'll simply put a spotlight out to illuminate your panels through the night to prevent theft but also with the idea the light shining on your panels will recharge your batteries through the night thus creating the first perpetual technological advance. If only it were that easy. Please don't do that since you will only create a dead battery by morning. As the sun slowly works its way up in the morning, your spotlight will be acting as though it were a sunset as it dims to a sad flicker.

The panels these days are very easy to wire since the connections can only go one way. Back in the old days, I had to make all connections in junction boxes; now the wires are all ready to connect together. A real good advancement.

If this was a permanent off-grid home or cabin, we would want the panels permanently and securely mounted either on a roof or ground mount. Roof mounted panels were our choice in Saskatchewan. It was not a good choice but it was really our only option. We were in the middle of the forest and we needed to situate the panels as high as we could. The roof fit the bill. It was a great location as long as it didn't snow.

Snow has a funny way of showing up in the winter when we need the power most. And northern Saskatchewan has a funny way of having winter show up in October and extend through April. A snow-covered metal roof on a 2-story building that is 100 miles from nowhere is a poor place to be, but I was up there countless times cleaning off those snow covered panels. To make the task as safe as possible, I mounted a wooden ladder affair to the roof to give me some footing and I always tied myself into a rope that was tethered to an anchor on the roof peak. I'll sum up by saying that snow covered panels are a major problem for roof mounted arrays. We now have a ground mounted array which is very easy to sweep off.

Our Nova Scotia Solar Array

Since this modular power system is meant to be a temporary solution to the grid down, the majority of our customers will not need to worry about permanently mounting panels. In order to make our book as useful for as many people as possible though, I'll assume there might be some who might wish to permanently mount their panels and I'll provide useful information for that possibility. Regardless, it would be best if you had a general idea of your latitude since you still want to try to capture as much solar energy as possible. The idea here is to carry a panel or panels outside if the power goes down, set them on the ground with a stand of some sort, and have them remain stable in an angled position. I could have bought fairly expensive mounts for the panels but I wanted to keep the cost as low as possible for you. By all means, you are free to purchase the commercial mounting frames to set your panels on the ground when you purchase your panels. But when it comes time to deal with the panels, I'll have a few suggestions you can do to set your panels up far cheaper.

Just like a greenhouse, solar panels are best when the sun's rays are directly beating down on the surface. But since the sun's angle changes throughout the year, unless you have a way to shift the angle of the panels each season, a compromise is in order. The rule of thumb is to set the panel angle to your latitude. As an example, we live at 45 degrees latitude. Our panels are mounted outside at the same tilt of 45 degrees to the sun. No need to break out the protractor. Just use your best judgment.

The easiest way to find your coordinates is to use the internet and search for your town or city and enter the search term "coordinates." Make sure you enter your state as well since it might spit out a town on another continent if you aren't specific.

There are some tweaks to the latitude rule but to keep it simple, let's say panels should face due south with an angle set to your latitude. At this point, you have a couple of decisions to make regarding positioning your panels. But the one rule is this. In order to make the most power from a solar panel, it must face the sun and the rays of the sun must beat straight down perpendicular to the panel. It might help to think of your solar panel as a mirror. You don't want to deflect the sunlight off the mirror. You ultimately want the sunlight to bounce straight back off the mirror.

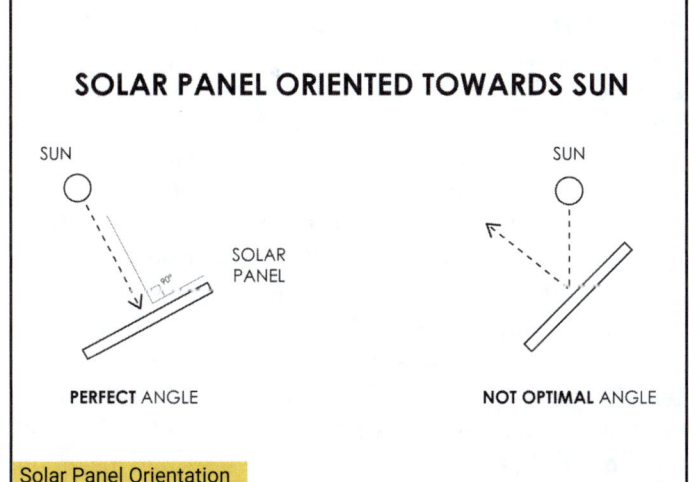
Solar Panel Orientation

Same with your solar panel. Because this is temporary and because you have complete flexibility if you are at home, you can arrange your panels and change the angles several times a day. But this requires you to remember to go out and reposition the panel/s.

You could set your panels in the morning directly towards the sun, reposition them direct to the sun at mid-day and then once again turn the panels towards the sun later in the afternoon, effectively mimicking a rudimentary manual solar tracker.

But you will not gain anything if you set the panel in the morning towards the sun and then forget about it so that the back of the panel is facing the sun in the afternoon. That defeats the whole purpose of trying to capture maximum solar energy.

If you have to leave the setup for any reason during the day, it's probably best to just set the panels up facing south and treat it more as a permanent mounted system. That way, it captures considerable energy without you having to go out multiple times a day to move them around. What you decide to do is purely your decision based on how much attention you want to give your system versus the need to capture the most daily energy.

The next question to ask is where in the sky is the sun located around noontime? Why? Because you want to face the panels towards the sun at that time. Doing so means you've essentially taken the seasonal variation in to account. In other words, because the sun's angles change not only through the day but through the seasons as well, regardless of when your power goes out, once your panel is set up at the correct angle around noon time, you should be good to go.

If by chance you want or need to use a compass to find due south, the following paragraphs will aid you. It's not as simple as taking a compass, orienting the needle to face south and voila, due south is located. Declination needs to be taken into account. We know our panels must face south for maximum energy production and if we don't take declination into account when locating the panels, we won't be getting maximum energy output. Here's a trick mentioned by my solar electronics dealer in the event you don't have a compass. Put a stick in the ground, mark the shadow every half hour from 10:00 AM to 2:00 PM. The shortest shadow denotes south.

If you do have a compass, hold it level in front of you and rotate your body until the needle points north. Usually, a red painted end will be the north end of the needle. Directly behind you is south. You can now reorient your body so that you are facing south. Double check the needle on the compass so that behind you, the needle still points north. This method gets you in the ball park and will usually be within plus/minus 20 degrees in North America.

In order to be more precise, for example when orienting your house, garden or a permanent solar array due south, you'll need to factor declination into the mix. Your compass is actually pointing to a drifting target caused by the interaction of earth's solid iron core and a molten outer layer of iron moving at a different speed. That action creates a magnetic field that is in a constant state of flux. If you remember the globes at school that rotate, true north is the top pivot point, the imaginary axis on which the world rotates. Depending on where you are on the earth's surface, magnetic north and true north will be different.

That drifting magnetic field and the axis pivot point at the north pole are in different places. The difference between the two is referred to as declination.

Here's a great site for figuring out your own declination. https://www.magnetic-declination.com/ Then it will be a matter of adding or subtracting the declination depending on if your declination is a positive or negative value.

For those who have no interest in or don't foresee the need to orient something precisely with a compass, don't bog down on the concept. There are many good books that can help you, perhaps a friend is good with a compass or as a last resort, a GPS with compass should take care of the declination automatically if you ever needed to be more precise.

In summary, remember the following and you should be OK. In order to make useful energy production, your panels must face south, be at the proper angle, your latitude, not be in any shade, and not be located in the closet. Let's move on to the Charge Controller.

Charge Controllers

The charge controller is the component that goes in between your solar panels and the batteries. It controls or regulates the energy produced by the solar panels which in turn goes to replenish your batteries. This is an electronic device which senses when the batteries are full or if they need more charging. Without some kind of regulator in the system, the solar panels will put all available energy to the batteries whether the batteries need it or not. And if the batteries are completely full, there's a potential of damaging those expensive batteries by overcharging and generating heat. At the very least, the power is simply boiling the battery electrolyte away. Not only that, but your system voltage will rise to dangerous levels.

In order to charge the battery, you must force a voltage higher than it's nominal voltage. For example, a 24V battery needs to have a voltage supplied that is higher than 24V. And as the battery takes on a charge and starts to get replenished, that voltage will rise and continue to rise as it approaches fully charged. Without some mechanism to regulate the charge, the voltage could easily go over 31 volts.

You may damage appliances and other electronics that are connected to that high voltage. You might have 24VDC devices that operate within a safe range of 24V. But once that range is exceeded, the device will either not work or may be destroyed by a voltage it wasn't designed to operate in. Or you may have devices with built in over-voltage protection and once the voltage exceeds a certain threshold, that device shuts down to protect itself. The inverter, another electronic device we will talk about next is an example with a shutdown feature. So, you can see, a charge controller is a must for the typical system to function without failure.

There are 2 different types of charge controllers: PWM and MPPT. For our purposes, we don't need to know how the controllers actually work. The MPPT is the more sophisticated choice and it is the type of charger I selected for our system. The MPPT charger is generally more efficient, better suited to lithium batteries and more versatile, but the downside is it's a little more expensive initially.

Both types of controllers have their use but what's more important is to know the theory on charging. We all have better things to do during a day than to monitor our batteries so they aren't overcharged and destroyed. We'll let the smart electronics do the monitoring and adjusting for us.

Here's what you need to know about charging. A good charge controller will have 4 stages of charging: bulk, absorb, float and equalization. The controllers are computer controlled but can be tweaked by the user if desired. Different battery types and manufacturers may specify or require certain charge parameters that are needed to best charge that particular battery.

Pretend you have a quart jar with a very slow leak. If you needed to precisely fill that quart jar with water to the top from the faucet without wasting water, you'd hold the jar under the faucet and turn the water on full. As the level rises closer to the top, you'd feather the tap so a slower stream comes out. When you got really close, you'd probably turn off the tap and give it a couple quick on-off sequences to just tweak a few extra drops right to the top of the quart rim. You've just manually simulated a bulk and absorb charge. As the quart jar slowly leaked drop by drop, you'd replace those drops lost with drops from the faucet. You are now in the float stage which is just providing enough drops to the quart jar to keep it full to the top.

That's the main function of the charge controller. Fill the quart jar (battery) and do it automatically in stages. A good controller has pre-programmed functions that monitor the voltages, current and time. One of the first things we want to do is let the controller know what type of batteries we have. We'll delve into that when we wire our system together. Once the controller knows what battery type, it will monitor the battery for specific parameters.

At the start of the day or when our panels are initially hooked up, and assuming our controller has been set up initially, our controller will look at the voltage of our battery. Based on the voltage it sees, it will select the proper charge sequence. Usually, it will start out in the bulk charge which allows the solar panels to put maximum power to the batteries. Full solar panel energy will be sent to the batteries until the controller senses that the batteries are getting full, then it will drop to the next stage which is absorb. Now the controller is feeding the batteries a current that is regulated and it's the process of gently topping off the batteries to a full state of charge. Then finally, the charger goes to the float mode which is just enough of a current to trickle charge and maintain the batteries so they remain fully charged. The equalize state is the last stage and is only done when necessary. We discussed when that is needed in the battery chat. Because we selected lithium batteries for our project, equalization will not be needed and we won't delve into any further.

Another consideration when buying a charge controller is to oversize it. For a little extra money, consider buying a controller that will handle double or triple the capacity of your panels. That way, it will be very easy to upgrade and install more panels without having to buy a whole new charge controller. Or at least inquire whether the charge controller you buy can be connected to another controller if you decide to expand your array size. I took care of this step for you when I selected the controller for our project. The controller can handle up to four 200-watt solar panels. That is the maximum power the system is designed for. I will mention that a second charge controller could be purchased and installed if one wanted more than 4 solar panels but, in my mind, that's beyond what a temporary power system should be to bridge a power outage gap. Who wants to be lugging more than four solar panels out to the front yard?

Inverter

The inverter is a piece of electronics that takes a DC (direct current) voltage such as our 24VDC battery voltage and converts it to 120VAC (alternating current), so we can function like any typical house. In Saskatchewan, where we had a generator, the inverter also sensed when the generator was running, and then the inverter worked in reverse so it became a powerful charger to replenish our batteries.

I made the choice of a specific inverter for our backyard power plant for many reasons and one of them was the simplicity of wiring. As I mentioned, more sophisticated inverters have the ability to be hardwired into a generator so that if the batteries were low, you could run a generator. That generator would run any appliances plugged into it while at the same time supply power to the inverter so the inverter acts as a battery charger to recharge the batteries. That would add unnecessary complexity and cost to the backyard power plant and make it more confusing for you to wire in. Neither was something I wanted to saddle you with.

Having said that, many people do own a generator and we can still utilize it to charge our batteries. Instead of an expensive inverter with a battery charge feature, we will use an external, separate charger we can hook up when needed. No different in concept than if you wished to recharge your car battery with a car battery charger.

I specified and purchased a battery charger made specifically to charge lithium batteries either from grid or generator power. We can use this charger initially to wake our batteries up at the time of purchase and occasionally keep them topped up while in storage when we have normal grid power available. The battery charger I specified is also a Plan B so that if for some reason you are unable to fully recharge your batteries using the solar panels/charge controller of our Modular Backyard Power Plant you can do so using a generator. The charger is a cheaper means to recharge batteries than purchasing a more expensive inverter. Just plug the battery charger into the generator and recharge your batteries without any need of complex wiring. It gives you more charging flexibility.

There are pure sine wave and modified sine wave inverters. Sine wave inverters used to be quite expensive. Nowadays, they have come way down in price and it makes the most sense to buy one of them in the power configuration you need. Our backyard power plant uses a pure sine wave inverter. Sine wave power, which is alternating AC current, is what the utility company produces. So, in essence, a radio or TV can't tell the difference between being connected to the utility company or an off-grid sine wave inverter. The 2 power sources are the same voltage and quality.

The modified sine wave inverters are what we used to use in the past since they were what we could afford. But now that they are more reasonably priced, our current Nova Scotia homestead is powered by a pure sine wave inverter. If you could look at the wave shape, instead of a nice rounded top and bottom to the wave as in a sine wave, the top and bottom of the modified sine wave are clipped; they are squared off at the top and bottom of the wave.

There might be a fussy piece of electronics that won't

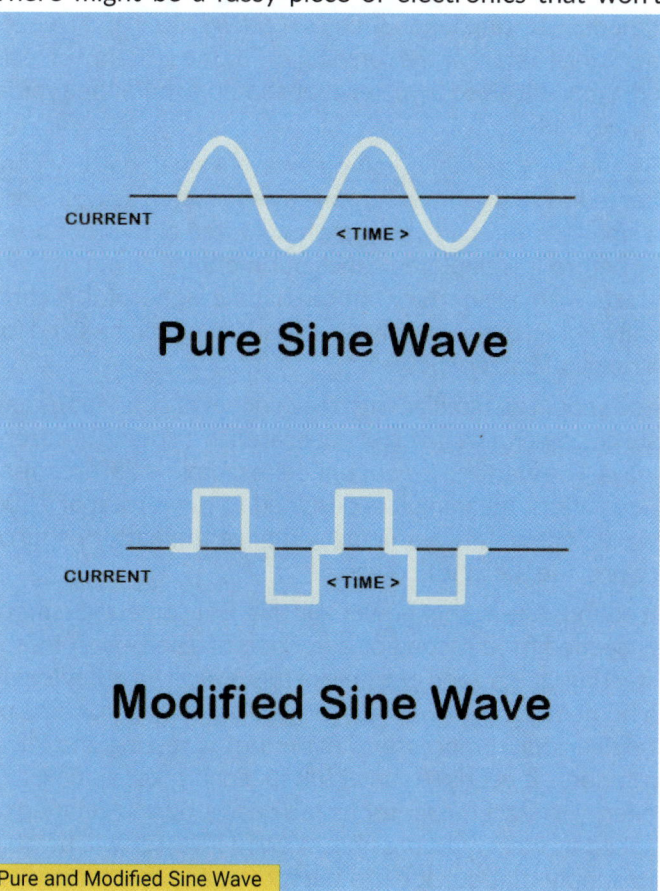

Pure and Modified Sine Wave

work well with that clipped wave but we never had any problems with it when our home used a modified sine wave inverter. We ran satellite modems, computers, TV etc off modified sine waves with no trouble.

Inverters come in various sizes depending on the intended use. When I use the term size, I'm talking about the wattage of the unit. How much can it power at one time. If you had a 200-watt inverter and expected to run a 1000 watt load, it won't work. It's only made to supply 200 watts. I selected a 2500-watt unit which is somewhat in the mid range of what is appropriate for a home. Unit selection is a balancing act. The more power an inverter is designed for, the bigger and heavier the physical unit will be. At some point though, it gets unwieldily for a temporary set up. But if too small, it won't power everything you need to run.

I had a hard time choosing an inverter size for the Modular Backyard Power Plant since many of you could probably get away with a 1500-watt inverter. It really depends on what you want to power up. I figured I'd oversize it a tad so folks would have the option of going for the complete system (Part-Time Power Plant) or for a larger battery bank than we designed for. As long as you have the battery reserves, the 2500-watt unit will give you a great deal of flexibility. I also didn't want to take any chances the system would go into fault mode during an emergency situation. That's the last headache you would want to deal with.

Much like the charge controller, I suggest spending the extra money for a larger inverter since you will likely add more and larger loads over time and you will have everything in place to handle it. It will be able to handle simultaneous starting loads such as a refrigerator and freezer deciding to start at the same time. For the system, I selected a modest sized inverter that should handle the typical home loads.

By the same token, if you only have a small battery connected to a large inverter, that's a mismatch as well since a small battery can only supply a limited amount of power before it is depleted. Since our modular power system starts with one battery but can handle eight total, technically, I needed to select an inverter that would be sized for a range of battery power.

You should be familiar with the concept of continuous and surge power. The inverters are rated in a couple different ways. The wattage for our unit for example is 2500 continuous watts, meaning it is designed to run a total of 2500 watts indefinitely assuming you have the battery source to provide the input power.

It sounds like a lot of power and it is. But sometimes, more is needed for a second or a fraction of a second. A motor starting up is a good example of this. We all have devices in a home that have a motor, perhaps a washing machine or refrigerator compressor. That motor is resting and all of a sudden, it needs to run at full speed. It takes some extra energy to start the motor from a dead stop and get it up to speed as quickly as possible. A good inverter will be able to supply that quick extra burst of energy without going into self-destruct mode. The inverter is designed to supply a spike of extra power and once the motor is up to speed, supply the needed energy to keep that motor running. Or perhaps two devices decide to start up at the same time requiring a big surge of energy. The inverter should be able to provide the energy for that brief period without shutting down.

I mentioned going into self-destruct mode. That's never good. Fortunately, in over four decades of using off-grid equipment, we've never had an electronics failure. Modern devices have built in protection features allowing them to protect themselves before any damage is done. For example, if a higher input voltage shows up, it should shut down. If it is over loaded by trying to run too many things all at once, it will shut itself off. If it gets too hot, there are sensors that will tell the inverter to turn off. My point is this equipment is reliable and has built in safety features. If for some reason, your house was running and all of a sudden, everything shuts off, it may be the inverter sensing something is wrong. Don't assume the inverter died. The inverter is giving you a warning and a direction to start troubleshooting. We'll have a troubleshooting section later in the book.

And finally, many modern inverters have several modes. There are times when there's no load to power. The refrigerator doesn't need to run yet, no light needs to be powered and no devices need to be recharged. It would be nice if our inverter could just go to sleep. It takes power to remain turned on and sometimes, it's nice to have the inverter go into a "standby" mode where it takes very little power to run. Then when a load needs to run, it comes out of power save mode and runs the appliance. Our inverter has such a feature which we'll discuss later.

We've touched on the term efficiency in previous chapters and we should touch on it again. Inverters are a really cool device to convert direct current to alternating current and sometimes do that in reverse. But there are losses to the process. That electronics box takes power to run itself regardless of if you have a load or not. In other words, if you turned all your devices off, the charge controller and inverter are still taking some battery energy. When we discuss the system design, we'll take efficiency into account.

Miscellaneous Components

In addition to the 4 main components of a solar electric system, there are a host of other very important pieces. Some of those pieces include a combiner box that takes the output of various solar panels and makes it easier to run a single cable to the house. We will not need one for our system since it isn't absolutely necessary and adds to system cost.

Bigger systems though need a combiner box. It's just an

electrical box that lets you combine the outputs of many solar panels right at the source instead of running many wires into the house and then having to deal with a mess of wires. Using a combiner box makes for a safe, clean installation when dealing with lots and lots of solar panels. Since we only have a maximum of 4 panels, we will bypass the combiner box and use a couple "Y" connectors to tie everything together.

Other electrical boxes and enclosures are an important piece of the overall safety of the system. Depending on the box use, they can house other components such as circuit breakers and fuses, measuring devices as well as a junction box, which is a safe place to join wires. It provides a way to keep all the wires in a neat, orderly arrangement. It prevents damage to the joint where wires are joined together.

Circuit breakers and fuses are mandatory. They are not only protection devices in case of a failure or short circuit but they are also a means to safely disconnect devices and loads. Much like wires, circuit breakers and fuses must be sized appropriately for the current they can handle or the current needed to trip them. You don't want the wire burning up before the breaker trips. Once a fuse is blown, it must be replaced. Circuit breakers are essentially the same devices you already have in your home in a breaker box. You will have many breakers in the box protecting various circuits in the house. And just like your home breakers, the breakers we have selected for our project are easily reset should they trip.

Be aware, there are circuit breakers made specifically for AC or DC circuits and any breakers need to be rated for not only voltage and current but also whether the voltage is AC or DC. For our DC circuits, I have specified the correct breakers. You couldn't run to the hardware store and just snag any breaker off the shelf. The breaker would need to be DC rated if used to control a DC device.

Wires and cables. We can't have a solar electric system without wires to move our energy from place to place. You already know how critical it is to use the right wire size. All the wires have been properly sized for our system but if you do anything additional with electricity, a visit to a wire gauge table will confirm you have chosen a safe wire for the task at hand.

Meters are not mandatory but are a nice addition to any solar electric system. They provide a way to visually see what the battery voltage is, how much the solar panels are producing, how much current a refrigerator or other load is consuming and what's the status of our batteries. Nice information to have but we can live without it. That would add significantly to our cost and since this is a temporary setup, all we really care about is keeping our appliances going until power is restored. Our charge controller will have a monitoring system to give us a reasonable idea of how well our system is working.

The last thing I'd like to mention for this chapter is this.

Much like buying a car, there are all kinds of manufacturers and models. Regardless of what make and model of car you buy, it essentially acts the same. You open the door, get in, turn the key and drive to your destination.

Although I have specifically selected components from a specific manufacturer for our project, there are dozens of components I could have selected. They all work essentially the same way and the wiring theory will be the same. Let's look at a battery for example. All batteries are going to have two terminals no matter where or who you buy them from. A plus (positive) and a minus (negative).

The electronic components to our system are based on the same principle. My decisions were based on a combination of specifications and price for what was widely available today. Just like cars are constantly changing each year, so too are the electronics. Although I took extra time to explain my thought process with this stuff, I did it so in the future, you can have the knowledge to select another product to do the job if for some reason what we selected is no longer available or another manufacturer has a better price. The wiring theory will be the same regardless of what manufacturer you select.

CHAPTER 5
Modular Backyard Power Plant Design Theory

Here's as good a place as any to set out what each module of our Modular Backyard Power Plant can power. Remember, we have the 3-Day Blackout Power Plant with one solar panel and one battery. Then you can add the One-week Blackout Power Plant module which is an additional battery and panel that will effectively double your energy output if you need to. And finally, you could add the third Part-Time Power Plant module which maxes out what the controller can handle. The Part-Time Power Plant module adds two more batteries and two more solar panels.

Let's take each module one at a time. Also please note... I'm not into hype. I have done actual calculations based on what typical appliances consume energy wise in a day. And I will be conservative in my estimates. We will talk a lot more about how I calculated and arrived at these conclusions. I will show you exactly how I arrived at my figures so you can do the same thing at any time with your appliances. Then you will have some realistic expectations for your new system.

Satisfaction with any off-grid system, whether permanent, or in our case temporary, is dependent on whether it is sized properly. For anybody who has plans to someday design a more permanent off-grid system, I urge you to have your design checked by a professional installer or a sales engineer from where you will be purchasing your system components. Have them double check your calculations and reasoning. Too much time and money will be spent on this and if things are not sized properly, you will either have spent more than was needed or you will be woefully underpowered.

As you continue reading this chapter, you will see I've done all the hard thinking for our Modular Power Plant. All you need are the components and a desire to wire it together.

3-Day Blackout Power Plant Module

I hope you will have some mercy on me. As someone tasked with designing a system to run your devices, I have to use experience and good judgment coupled with typical specifications of gadgets and appliances in order to calculate energy consumption. Total energy consumption of a device is an important consideration.

If you've ever gone shopping for an appliance, it will usually have an energy rating. It uses "X" amount of energy in kWh/year. Most people have no need to understand what that means. All that is important when shopping for an appliance is making a comparison between several similar sized units by looking only at the tag and choosing the unit that uses the least amount of energy. But you aren't like most people. You will come to see how important it is to give energy consumption critical consideration.

There are so many variables. Let me use a "refrigerator" as an example. Is it energy star rated meaning it has been tested for efficiency and meets certain requirements? What is the size of this refrigerator? Apartment sized or jumbo deluxe? Does it have a freezer compartment?

Is your location always warm to hot so that the refrigerator runs more than average? Are you or family members constantly opening the door to grab an item? And if so, are we casual about it so that cold air is just dumping out of the open refrigerator into the warm room? That refrigerator has an adjustable setting. Is the adjustable temperature setting set for the typical refrigerator or is it set to the "I want my beer with ice crystals in it" cold. Or put another way, how long is this thing running in a 24-hour period?

Randomly pick 10 neighbors in your neighborhood and inquire about what size, make and model refrigerator they have and I'll bet you get 10 completely different refrigerators. And yet, in order to design a system and have confidence in what it will power up, I or any other system designer needs to use some reasonable estimates of a "typical" refrigerator in order to come up with a proper design.

I used spec sheets for any appliances when possible. That way, we can calculate very close to what the energy demand is for each appliance we want to power. Also note, I am making the call on what I consider the essential appliances you must run if the grid fails. I am putting myself in your position as I design this. I will justify why I selected certain priorities. These are things I would consider the most important and that's what we will use in our system design. You may disagree and that's fine. You are free to substitute as long as you understand the power requirements of your substitution.

I found it interesting when researching the competition that there are some wild claims on what the competition's system can power and for how long. If people only understood the balance needed between what they want to power and the capacity of the system they purchased, they'd realize in a true prolonged grid down scenario, their system would be nothing but a source of frustration unless it was properly designed to handle the priority appliances. The battery of the competing systems would be dead in short order plus many of the competing systems have no way to expand.

You are now getting a sense of my thought process so let's walk through the three modular packages and conservatively estimate what they can run and for how long.

But before we get there... I have to tell you, I've been sitting here pecking away for you, my dear reader. Trying my best to give you the the most comprehensive book on the market with all the information you need for an off-grid power plant. You have been my sole focus. When I sit here typing away, I am giving you 100% devotion.

Yet, in all honesty, Johanna will come in and look over my shoulder and take me off stride, or sometimes, she'll ask me to give her a hand with something or perhaps I just need to get up and stretch. I have the best of intentions to write for an hour nonstop, but something always seems to take attention away from the task at hand. Although I sat down to write for 60 minutes, I was only able to get 51 minutes of actual writing time in. I lost 9 minutes of writing time to disruptions. Are you confused?

The point is I'm not really 100% efficient with my writing time because of the disruptions. I wasn't as efficient with my time as I had hoped! Nine minutes were lost out of that hour to other things.

The same principle applies to our calculations. Efficiency is extremely important in our calculations. I mentioned efficiency when I used the refrigerator as an example. Everything will have losses that must be factored in. If we are to do a good job designing our system, we need to understand the term efficiency. I will tackle it in upcoming examples.

I want to define what I consider the priority appliances: refrigerator, water pump, smart phone, light, medical device. Here's the reasoning. The refrigerator hopefully has a freezer compartment. The unit is the device we need to run to keep perishable food from going bad. Assuming grocery stores are available within reasonable driving range, a refrigerator will give you the ability to stock up for a period of time. I can't know and predict every circumstance where our power plant will be useful but a treat in the form of ice cream or a cold soda/beer will keep the spirits up while you wait for the restoration of power. Since most people have a refrigerator full of food, it makes sense that keeping the food safe is a priority. We do not want that food going to waste! In a worst-case scenario, that food will be a much needed source of energy.

Having a freezer of some kind also gives you the ability to freeze water into ice blocks. Those ice blocks can be used in multiple ways, one of which is to keep food refrigerated using a simple ice chest. Many homeowners have a cooler for camping and sporting events and now is the time to consider utilizing it until the crisis is over. I talk about this more in Chapter 11.

In much the same way as food is essential, water is vital for survival. I can't know everybody's source for water so all I can do is plan for the potential that some of you will need to pump water in some fashion. If you have a source of water that does not need an electric pump, that's bonus energy savings that won't be expended for water pumping that we can utilize elsewhere.

We've never been hooked to public water so I can't talk intelligently about how public water will work for you if power goes down. If in doubt, inventory water. At the very least, have a plan to provide drinking water.

Our experience over 43 years has been with drilled wells, machine dug wells, hand dug wells and sucking direct from the lake to provide our water. Water taken from any open source such as a river or lake must be treated with filtration and/or boiled to make it safe.

I have to be honest, securing water in a grid down situation is problematic and I urge you to think this through before power goes off and you are left scrambling.

Here are a couple of problems. Generally, the water pump for a home is direct wired into the circuit breaker box. It is on its own circuit. There is no handy plug you can just unplug from the wall outlet and then plug into our Modular Power Plant.

The next problem is the voltage. Some deep well pumps are made for 220V. Our system will not handle 220V. There's no way I can provide a water solution for everybody given that there are different well types, different plumbing and different voltage and overall size configurations. However, the following are some points to consider.

If you have access to a shallow well, you can easily have a standby shallow well pump that can be plugged into our Modular Backyard Power Plant. You also have the option of running a 24VDC pump such as the type used by Rv'ers. However, here's another potential problem to be aware of. Many pumps have a pressure switch. It's a means of controlling the pump.

When the pump comes on initially, it pumps water into the house and pressure tank. As the system fills with water, the pressure starts to rise. We can't allow the pressure to continue rising until something breaks. We need a way to sense that pressure so when it reaches a safe set point the pump automatically turns off. Hence the need for a pressure switch. As you use water in the house, the pressure drops. When the pressure switch senses the pressure is too low, it turns the pump on again and this cycle repeats as water is used in your house.

The problem you may encounter is this pressure switch is controlled by a small coil that doesn't consume much energy to function. If you have your inverter in power save mode, the small amount of energy needed to energize the pressure switch coil may not be enough to bring the inverter out of sleep mode. In that event, wait for a load to bring the inverter out of power save mode, such as the refrigerator coming on. Or plug a device in that gets the inverter running and then pump water at the same time.

Or another thing you can do is reset the inverter to be on full time, pump your water and then reset the inverter back to power save mode. It's only a matter of flipping the right DIP switch on the inverter, powering down and then powering back up.

The alternative is to temporarily bypass the pressure switch and power the water pump motor directly. However… YOU MUST KNOW WHAT YOU ARE DOING HERE! You cannot wire the pump up direct to the motor and just walk away. Fill up your buckets or containers and then turn the pump off. Otherwise, you have bypassed the safety pressure switch and as long as there is a water source and the pump is running, water is being pumped. Thus, we do not want to damage the house plumbing or create a flood.

If in doubt consulting with a plumber or electrician would be my best advice. Let them provide the best solution for your particular situation and how best to utilize your power plant during a grid down situation.

The smart phone is a means of communications. It is also a source for news, weather and updates. Being in touch with family and friends, being able to share news and let others know you are well will provide a measure of well-being and comfort.

If you've ever been camping, you know the campfire was not only a source of light but a source of contentment. It was a measure of reassurance which kept the darkness at bay and kept the general atmosphere of the situation positive. Having a light on at night will be the equivalent of a campfire. Something bright to latch on to give a sense everything's OK. Not normal but good until the power is restored.

Medical devices speak for themselves. If you have any electrical device needed for your health, well-being and safety, that's a priority we need to keep powered.

We now know what we want to power as priorities in a grid down scenario. And we know we want to deploy a solar power plant that uses solar panels to recharge our batteries. As you know, we have four seasons and for some of us,

THE MODULAR BACKYARD POWER PLANT

those seasonal changes come with dramatic changes in weather and amounts of sunlight. June has the longest days of the year. December has the shortest days. If you have some way to schedule your off-grid disaster around the June time frame (lucky you), you should have a good 6+ hours of peak solar charging. But the reality is the random disaster comes out of the blue at the most unexpected time and could very well coincide with the less than optimum time of year for solar charging.

The sun's height and the seasons

Seasonal Solar Variation

Looking at this solar map, all you need to understand is the United States is a big place.

The different shades of color are coded to show how much solar energy is striking the earth in different locations. Looking at all the color variation tells you there sure is a big difference in the amount of energy the sun provides at any given location. This map also tells us how many hours each day on average throughout the year we can expect peak solar charging. Peak solar simply means the sun is at such an angle that the energy falling on your solar panels gives you full output.

For you curious tech people, peak solar is a measurement of solar energy in kilowatt/hours per square meter falling on the earth's surface. And from that number, one can extrapolate peak sun hours. 1kWh (1000-watt hours) equals 1 peak solar hour. That's how I arrived at 4.5 hours of peak solar power for our calculations. The map shows most of the United States land mass receiving 4.5-4.9 kWh/m²/day. Since I have no idea where you live and there is such a wide variation across the country, we will use a somewhat worse case figure of 4.5 hours. It is a number that fairly covers the majority of the United States land mass and our customers.

What that number reflects is 4.5 hours of peak sun daily averaged throughout the year. It accounts for the seasons when winter days get shorter and summer days are longer. It takes into account cloud cover. It is an average accumulated measure of the amount of energy from the sun

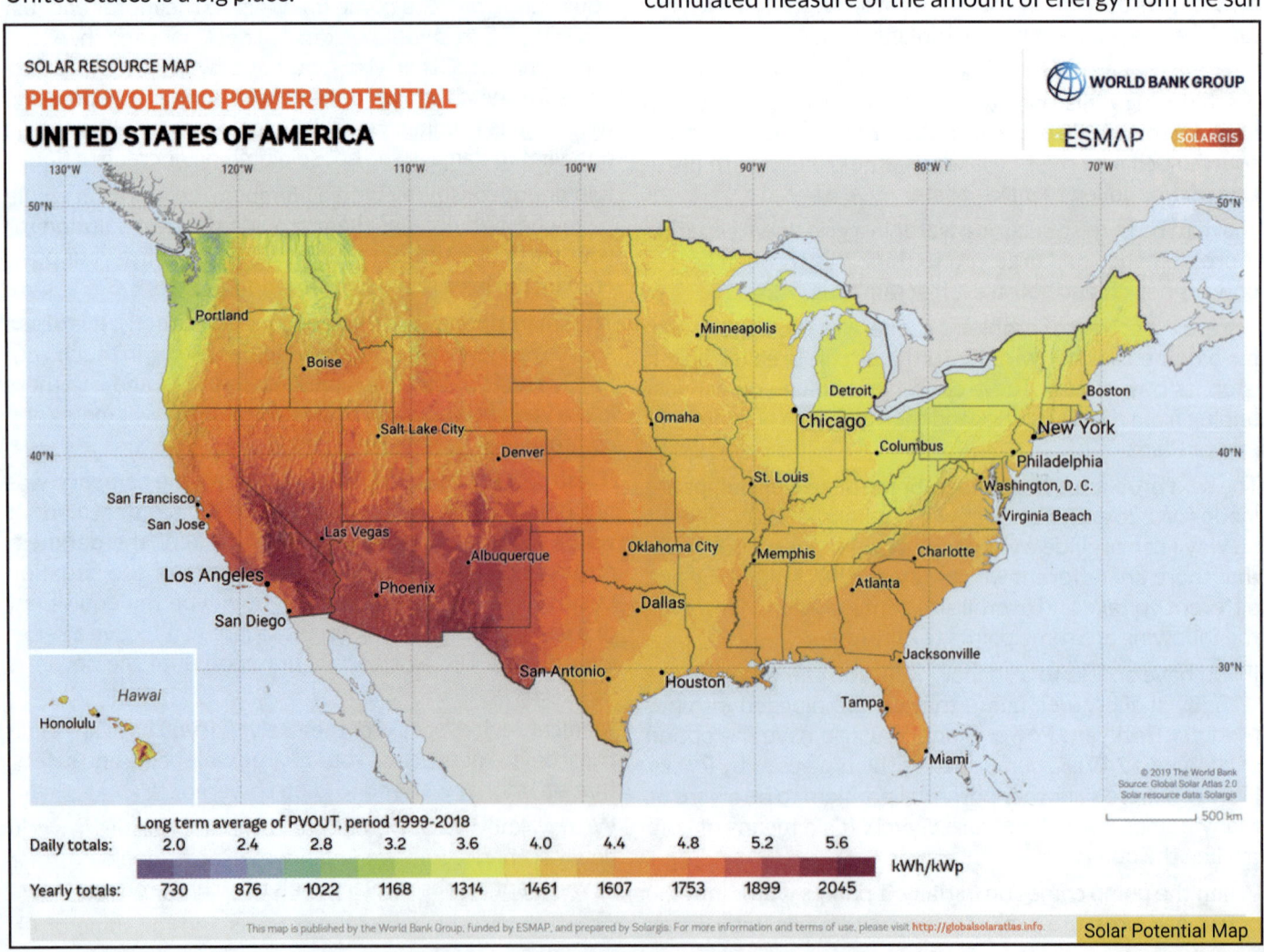

Solar Potential Map

from sunrise to sunset throughout the year.

Obviously, there are large swaths of the country that get more than 4.5 hours of peak sun. If you live in those areas, that means you will have a much easier time charging your batteries from the sun. Your system will work even better for you than what we have designed since I'm designing around more of a worst-case scenario.

Remember peak solar simply means the sun is at such an angle that the energy falling on your solar panels gives you full output. For our 200-watt panel, we are designing our system with the expectation we will have an average of 4.5 hours of peak sun every day throughout the year where our panel actually puts out the full rated 200 watts.

You can see how important that hourly number is. We need to count on those hours of sunlight each and every day to charge our battery bank back up. Keep in mind, your solar panels are still producing in hours where the sun isn't directly overhead. It's just that they are not producing the full output. Where you would collect 200 watts of energy from an hour of direct peak sun overhead, it might take 3 hours to collect those 200 watts starting from when the sun is rising onward to mid-morning and dropping down in mid-afternoon onward to sunset. Even on a cloudy day, your panel will be producing something and every bit helps.

There are several caveats worth mentioning. As we discussed under the topic of solar panels in a previous chapter, in order to get anywhere close to rated output, the panels must be in direct sunlight. You should know that when a manufacturer rates their panels, it's under perfect lab conditions. Rarely will we ever have full rated power in a day. We have to figure on something less.

Remember, in order to get maximum output from our panels, they need to be **facing the sun at the proper angle throughout the day**. The only problem is the sun is a moving target. Solar panels can be mounted on devices called solar trackers that follow the sun across the sky throughout the day but since our system is generally only deployed when the grid goes down, solar trackers aren't realistic for this application.

As well, we need to bring efficiency back into the discussion. There will be some loss of power through the wires, the solar charge controller, inverter and battery. Figure we lose 20%. But we will need to factor in even more loss when our inverter sits there waiting for a load to show up. We talked about that specifically in the Components Chapter 4. We will need to account for that loss in our designs below.

So, here's what we know now about our 3-Day Blackout Power Plant Module and design specifications:

- 4.5 hours of peak solar each day regardless of season or location.
- a single 200-watt panel which theoretically will produce 900 watts over the course of the 4.5 hours of peak sun. Realistically with 20% loss (900 X .2 = 180 watts), we have 720 watts of real power going into the battery daily.
- Our Renogy battery per the spec sheet contains 1280Wh of energy if fully drained. That's in theory. There are many factors which will determine the real amount of energy in the battery. One of the biggest factors is how fast that battery is drained. The bigger the load, the more stress on the battery and the less capacity the battery actually has in the real world. We really want to try to maintain 20% reserve in the battery.
- We know the battery has 1280Wh capacity. We know we have 720Wh of solar to recharge the battery each day. Can you see the problem?

If we deplete the battery more than 720 watts each day, we will never be able to fully charge the battery. In other words, let's say I run appliances and at the end of a 24-hour period, I've drained 1000 watt hours of energy. As soon as the sun comes up, I'm charging the battery back up but at the end of the day, assuming it was completely sunny and I shut all appliances off so all energy from the sun is directed back to the batteries, I made the maximum 720-watt hours to recharge the battery. Yet I'm still short 280-watt hours. Every day that goes by is accumulating shortfalls of solar energy to the battery bank. I'll never be able to fully recharge the battery. Our battery will be completely dead if we try to take 1000-watt hours out of it daily and only put back 720 each day.

Now, with those bullet points above as our starting place, we are ready to dig in and make some sense of what our starter package can do. Recall our priority appliances are: refrigerator, water pump, smart phone, light, medical device. I've researched the typical energy consumption of the priority appliances and summarized the information below.

You will likely have different appliances so you can look at the tag to see what the power need is for your particular device. Remember the equation back in Chapter 3, Basic Electricity? $P=EI$ If you look on the label and it has volts and amps, you can just multiply the voltage (E) times the current (I) to arrive at power. Then you can tailor your power requirements to your own needs and figure out what your own system is able to reasonably power.

Refrigerator – Come in all shapes and sizes with varying energy requirements. Generally, the bigger the refrigerator, the more energy needed to chill the volume of interior space. They may have a bottom freezer compartment or a top mounted freezer or just be a simple refrigerator. Our 3-Day Blackout Power Plant Module will realistically only be able to power a small refrigerator. You may see other portable solar power systems touting they can run full size refrigerators etc but what is not mentioned is the relatively short duration of run time. The numbers don't lie. At least you will now be able to evaluate any power system and the appliances it can run armed with all this knowledge.

Realistically, when I say small, I'm looking at a 5-6 cubic foot refrigerator which uses something like 575Wh/day.

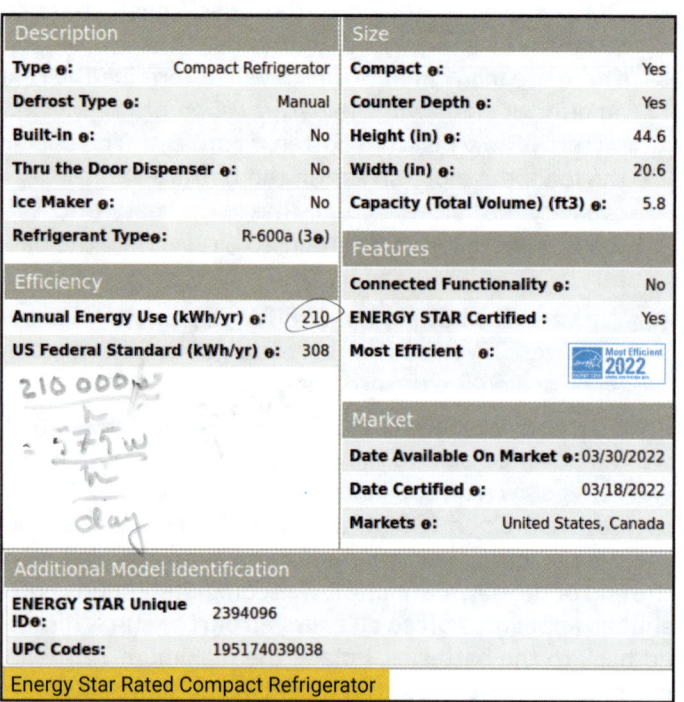

Energy Star Rated Compact Refrigerator

Note: I make no claims for the reliability or suitability of any refrigerator used in the following examples. They are merely proof of products and energy usage in order to show how I went about making calculations.

Our panel can only put out 720 Wh so our total appliance load should match. All I did was go to the energystar website and do a search for compact refrigerators. https://www.energystar.gov There are many choices.

Smart Phone - Johanna and I are still living in the stone age when it comes to smart phones. Friends and family have them but we have no real-world experience with them. We have a flip phone with a 5-watt charger. In researching smart phones, like everything else, I discovered they come in an array of models with charging capability of 5 watt to 20 watt typical. Judgment again comes into play. I don't know how long your particular phone lasts on one charge. I am going to make the following assumptions. A phone charge lasts 20 hours. The phone will be turned on for 5 hours a day. Charging through a USB cable takes 12 watts. A full charge takes 2 hours. Since we are only discharging the battery 25% daily, for the sake of our calculations, we will plug the phone in to recharge for 1 hour to top off the battery each day. That will consume 12-watt hours every day.

Water Pump – Another appliance with a wide range of sizes and power requirements. Before I forget, towards the end of this book I'll have suggestions on alternatives to some of these appliances. There are old ways of doing things that in a pinch could be valuable for you. As well, this water pump is best utilized if you can run it once, get all your water needs fulfilled and then shut it off. It's more efficient that way as opposed to start/stop/start/stop all the time. Remember we are after as much efficiency as possible.

A water pump is not on for long but it takes a lot of energy to push or pull water. Water conservation is paramount. In researching the power requirements of water pumps, it looks like 700 watts is a reasonable number. Keep in mind, as mentioned previously, many of you will have alternative sources of water. I have no idea if public water still works for you when the grid is down. Also, our system is only capable of powering a 110-volt pump. If your pump is 220VAC, you will need to address how you acquire water before the grid goes down. Something as simple as inventorying some could be the solution.

We will figure the pump being on for 5 minutes a day. A 1/3 Hp pump is capable of pumping 4gpm (gallons per minute) or more. Let's figure on 5 gpm and 5 minutes of running time. That's .083 of an hour. Our pump will consume 58 Wh per day. (.083 X 700 watts = 58 watts) Obviously, if you have a source of water not requiring an energy source to obtain, the energy savings could be used elsewhere.

CPAP Medical Device – If you don't need to power any medical device, then any energy savings could be used elsewhere. But these devices take 30-60 watts to run based on research from a dedicated cpap website. Let's go with 50 watts for an eight-hour night's sleep. That works out to 400 Wh of daily energy consumed.

Lights – If you don't already have energy efficient lighting, I'd suggest making the conversion. Not only will they help with the normal monthly power bills, but in a grid down scenario, they will provide much needed light without a serious drain on the batteries. I'd suggest LED lights. Look for the highest lumens for the power used. For example, a 9-watt LED provides the same light as a 60 watt incandescent bulb. We will figure on one 9-watt LED light used an average of 4 hours a night for a total of 36Wh (9 X 4 = 36Wh).

System Electronics – This is where we need to account for the power used by our inverter when it is sitting there but the refrigerator is not running and there are no other loads turned on. Our inverter, according to the spec sheet takes 1.4A/hr at idle which equates to 35Wh. Let's figure 12 hours our system is at idle with no loads. A load would be any of our appliances. That amounts to a whopping 420wh (35Wh X 12 hours = 420Wh) of wasted power with the inverter just sitting there powered up in normal mode.

Remember, our inverter has some special settings which we can utilize to drop that amount of wasted power. We can go into power saving mode. The only setting that makes sense for us is to use the 100-watt threshold. It's the lowest power saving threshold our inverter has. It's not an ideal situation but helps a lot. That power saving mode only uses .4A/hr which equates to a little more reasonable 10Wh. With our 12 hours of idle a day, that works out to 120Wh of wasted power.

The reason it's not ideal is now that we are in power save mode, the inverter looks for a 100-watt threshold to wake up or go back to sleep. Big problem. Let's say it's night time and we wish to be cheered by our bright 9 watt light. If the inverter doesn't wake up until it sees a 100 watt load, you will remain in the dark because a 9-watt bulb is less than the 100-watt threshold.

By the same token, maybe you got lucky and the refrigerator was running when you turned on the light. However, as soon as the refrigerator clicks off, it's bedtime because your light just went out. As soon as the refrigerator turned off, the load dropped below the 100-watt threshold and the inverter went into power saving mode. More expensive inverters have a "search" mode which uses very little power when it is idling. It's a wonderful feature but one that I could not justify the cost for you. If this was a permanent installation, that would be different.

Don't worry though, I have a solution which we'll go into with more detail when it's time to wire our system together. Basically, we'll bypass the inverter and go with DC right off the battery to power the low power devices. The numbers we are hashing out here will be valid numbers for our design purpose.

Here's what this tally will look like in graphic form.

Appliance/Device	Quantity	Watts	Hours Used	Daily Wh Used
Refrigerator (5.8 ft³)	1	72*	8*	575
Smart Phone	1	12	1	12
Water Pump	1	700	.08 (5 minutes)	58
CPAP Machine	1	50	8	400
Lights	1	9	4	36
System Electronics Normal	INVERTER ON	35	12	420
System Electronics Power Save	INVERTER POWER SAVE	10	12	120

*I am guessing on wattage consumed and run time. But the 575 Wh is the needed number. *Refrig usage/day*

We can now tally the daily watt hours needed to power all 5 devices for the day. This tally of 1501Wh of energy would be with the inverter in normal (always on) mode. If we had the inverter in power save mode, we would consume 1201Wh. If you needed to power all 5 of the devices listed, the starter system will not work for you regardless of inverter mode. You would need to upgrade and add a module. The numbers don't lie. Those are the numbers we have to work with. I can't know how many of our treasured readers need the CPAP machine, or need the water pump running. All I can do is make the calculations for you, explain how I arrived at our figures and each customer needs to evaluate based on their own needs.

From the above table, there's no way we can justify having the inverter in the normal, always on, mode. Our solar panel makes 720 Wh a day and our inverter is going to take 420Wh of that power doing nothing but waiting for an appliance to kick in. Unacceptable. So, from this point forward, we will assume the inverter has been set to power save mode. With that said, I am comfortable powering the small refrigerator, smart phone and light with the 3-Day Blackout Power Plant Module. That's 743Wh used each day.

Some of you astute readers might catch the fact that our load comes to 743Wh and our solar panels only put out 720Wh. The numbers are close enough. Your refrigerator might be a bit more efficient or perhaps the inverter takes just a fraction less current than the spec says. There will be many customers in locations where this will work and yes, some where this will not work.

Keep in mind, if you have more daily sun than we figured on, you can have even more confidence this should work. A half hour of extra sunshine in a day makes a big difference. The only way to know for sure is once the system is built, try it with your own appliances before a power emergency. Monitor your system. Learn when it is close to being fully charged or is at full charge. At that point, energy is just being wasted assuming the sun is still shining. Your batteries are full and you might as well use that energy some way. It's free! Turn on the phone again and have a chat with a friend.

One thing we haven't gone over is battery reserve or put another way, what do we do if the sun doesn't shine? Along with allowing for the poor timing of a grid down situation, we have to figure in and allow for the occasional day of clouds or rain. What then?

Our battery contains 1280Wh of capacity. I'd like to always leave 20% and never drain the battery completely to zero. That's 256Wh (1280 * .2 = 256) always left in the battery. That gives us 1024Wh of usable capacity. The 3 appliances I was comfortable powering with the 3-Day Blackout Power Plant Module consume 743Wh leaving 281Wh of reserve, roughly half a day's worth.

You know your weather patterns the best. If you live in the southwest US, you probably have day after day of sun. Conversely, if you live in the northeast, you might be faced with just as much cloud as sun. Everywhere else, somewhere in between.

There's only three ways to deal with a potential energy shortfall during cloudy spells. Shut all the appliances off completely, cut back and ration the energy to various devices or opt for more charging and battery capacity. We'll deal with those choices next.

THE MODULAR BACKYARD POWER PLANT

One-Week Blackout Power Plant Module

The above was a good exercise and evaluation. However, I think you'd agree that shutting off all appliances due to an energy shortfall is a really bad choice as is rationing power to the priority device. A better option is to move on to adding a second module. Adding a panel and battery is really the only viable solution. And it's so easy to do. Adding a second module, the One-Week Blackout Power Plant Module, consists of a second solar panel and lithium battery. We are now doubling our system. Instead of 200 watts of solar energy, we will have 400 watts. Our battery capacity will be doubled from 50Ah to 100Ah for a total of 2560Wh.

At this point, you have decisions to make and some thinking to do. Based on the above information regarding the 3-Day Blackout Power Plant Module, you now have a better idea of what one panel and one battery can do for you. Now that you've doubled your power with the One-Week Blackout Power Plant Module, your choices are to use the additional power of the second module purely as extra reserve to power the same items you ran in the above scenario or you can power some additional items or perhaps a bigger refrigerator.

In the interest of keeping this a fair comparison and easy to follow, I'll use our base case appliances again and we'll run some new numbers.

Here's what we know with the addition of another 200-watt panel and 50 AH battery to our 3-Day Blackout Power Plant Module:

- We still have 4.5 hours of peak solar each day regardless of season or location.
- We now have two 200-watt panels which theoretically will produce 1800 watts over the course of the 4.5 hours of peak sun. Realistically with 20% loss (1800 X .2 = 360 watts), we have 1440 watts of real power going into the battery daily.
- Our Renogy batteries per the spec sheet each contain 1280Wh of energy if fully drained. The two batteries wired in parallel now provides us with 2560Wh of stored energy. Again, that's in theory. There are many factors which will determine the real amount of useful energy in the batteries. The more stress on the battery, the less capacity the battery actually has in the real world. We still want to try to maintain 20% in the battery and not drain it down to zero.
- We know the battery bank has 2560Wh capacity. We know we have 1440Wh of solar to recharge the battery each day. We still have the problem that if we drained the battery completely, even with full sun, it would take at least 2 days to recharge the battery and likely a third day assuming we continue to power at least some of the appliances. But we have a lot more flexibility by adding the second module.

Appliance/Device	Quantity	Watts	Hours Used	Daily Wh Used
Refrigerator (5.8 ft³)	1	72*	8*	575
Smart Phone	1	12	1	12
Water Pump	1	700	.08 (5 minutes)	58
CPAP Machine	1	50	8	400
Lights	1	9	4	36
System Electronics Power Save* (Inverter)		10	12	120*

* Note that I got rid of the inverter in normal operation mode. Power save mode is the only thing that makes sense here.

With the basic 3-Day Blackout Power Plant Module, I was comfortable powering the refrigerator, smart phone and lights which used 743Wh each day. But we were going day to day dependent on sun shining for the full 4.5 hours of peak sun daily. We had half a day in battery reserves for a rainy day.

Assuming for a moment those are the only loads we wish to power, with the addition of the second module, we have 2560Wh of capacity. Again, I'd like to always leave 20% in the battery and never drain it completely to zero. That would be 512Wh (2560 * .2 = 512) always left in the battery. That gives us 2048Wh of usable capacity. The 3 appliances I was comfortable powering with the 3-Day Blackout Power Plant consume 743Wh leaving 1305Wh of reserve, almost two days' worth.

With our 3-Day Blackout Power Plant Module, there was no way we could power all five priority appliances so I had to select three to use in our example. Now that we have more power available, let's reexamine trying to power all five loads.

Our five appliances take 1201Wh. That is including the power used by the inverter. We have 2048Wh available so we can easily power all five loads if we had to using the One-Week Blackout Power Plant Module. And you have almost an extra day of battery reserve in the event of a rainy day. Don't forget, if you don't have a CPAP, obviously, the power savings can then be used for other things. Same for a water pump. Everybody is going to have a slightly different situation and there's an unlimited number of combinations you can put together as long as you understand the energy consumption of the item/items that you want to power.

For example, as an alternative, let's go back to the original three appliances, the refrigerator, smart phone and light.

The refrigerator was small. Let's get a much bigger refrigerator. How about 14.5 ft³ instead of a 5.8 ft³ unit. I can buy a 14.5 ft³ which consumes 923Wh of energy.

The Part-Time Power Plant Module

Solar Panel 200W each (handwritten)

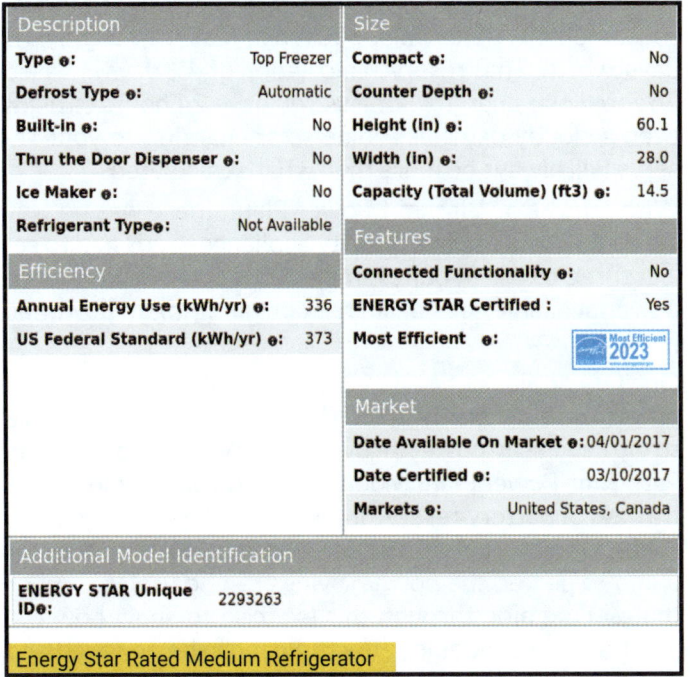

Energy Star Rated Medium Refrigerator

Appliance/Device	Quantity	Watts	Hours Used	Daily Wh Used
Refrigerator (14.6 ft³)	1	115*	8*	923
Smart Phone	1	12	1	12
Water Pump	1	700	.08 (5 minutes)	58
CPAP Machine	1	50	8	400
Lights	1	9	4	36
System Electronics Power Save *(Inverter — handwritten)*		10	12	120

* I am guessing on wattage consumed and run time. But the 923Wh is the needed number.

We continue to use the same table of appliances but this time, I've made the refrigerator a bigger model. From the above table with the larger refrigerator added, we have an energy demand of 971Wh. We have 2048Wh of usable battery capacity so we can easily power the refrigerator, phone and light and have over a day's worth of reserve left over. That should work fine. As you can see, we can power a lot with 2 solar panels and 2 batteries.

This third module, The Part-Time Power Plant Module will max out our solar charger capability. Adding this third module is as easy as it was adding the second module. The third module consists of 2 more solar panels and 2 more batteries. We are now doubling our system once again. Instead of 200 watts of solar energy which we had in our starter module, we will now have 800 watts. Our battery capacity which was 50Ah in our starter module will now be 200Ah. And we will now have a total of 5120Wh of battery capacity. You know the basic routine by now. You have a good idea of what one panel and one battery can power with the 3-Day Blackout Power Plant and you know what is possible with the addition of another solar panel and battery. After adding the extra panels and batteries of the Part-Time Power Plant Module, your choices are the same. You can use the additional power of this third module purely as extra reserve to power the same items we ran in the above scenarios or we can power some additional items and increase the size of the refrigerator.

Back in Chapter one, I mentioned the highlight of our off-grid life, the privilege of living on a remote lake in the wilderness of northern Saskatchewan. On top of the roof, I had an 800-watt array mounted. Our home was a 2-story building and if you flew out for a visit, other than a composting toilet and seeing solar panels and a small wind turbine, you'd never know you were in an off-grid home.

We had two chest freezers, refrigerator/freezer, lights, satellite TV, satellite internet, desk top computer, water pump etc etc. Johanna had her kitchen Aide mixer and other kitchen appliances to keep us well powered with chocolate goodies and other treats. The solar panels from about March to October pretty much powered the house with no problems other than prolonged cloudy spells. The 1000-watt wind turbine was supplemental power during winter and cloudy periods. A rarely used small 6000-watt generator was the third backup option. We took advantage of the cold climate by setting our two chest freezers outside under a protective porch so that in the winter, they were unplugged. No good reason to have freezers plugged in when the temperature outside is minus 20°F. I'll have a bunch of suggestions at the end of this book on how you can use the situations you are in to your advantage.

But my point is you will have the same solar power with The Part-Time Power Plant Module as we had in our full-size remote home. The only difference is we had a much larger battery bank. We had a 2000AH battery capacity. Where as in our Modular Backyard Power Plant we have 200AH available with The Part-Time Power Plant Module.

Let's run some new numbers for The Part-Time Power Plant Module using our base case appliances again.

THE MODULAR BACKYARD POWER PLANT

Here's what we know with the addition of another two 200-watt panels and two more 50 AH batteries:

- We still have 4.5 hours of peak solar each day regardless of season or location.
- We now have four 200-watt panels which theoretically will produce 3600 watts over the course of the 4.5 hours of peak sun. Realistically with 20% loss (3600 X .2 = 720 watts), we have 2880 watts of real power going into the battery daily.
- Our Renogy batteries per the spec sheet each contain 1280Wh of energy if fully drained. The four batteries wired in parallel now provides us with 5120Wh of stored energy. Again, that's in theory. We still want to try to maintain 20% in the battery and not drain it to zero. And a heavy load will still affect overall capacity.
- We know the battery has 5120Wh capacity. We know we have 2880Wh of solar to recharge the battery each day. Depending on how far we discharge our batteries, it still may take several days to fully recharge the batteries.

Appliance/Device	Quantity	Watts	Hours Used	Daily Wh Used
Refrigerator	1	72*	8*	575
Smart Phone	1	12	1	12
Water Pump	1	700	.08 (5 minutes)	58
CPAP Machine	1	50	8	400
Lights	1	9	4	36
System Electronics Normal		35	12	420
System Electronics Power Save		10	12	120

With the basic 3-Day Blackout Power Plant, I was comfortable powering a small 5.8 cu³ refrigerator, smart phone and lights which used 743Wh each day. Assuming those are still the only loads we wish to power with the the full capacity of our power plant, we have 5120Wh available. Again, I'd like to always leave 20% and never drain the battery completely to zero. That works out to 1024Wh (5120 * .2 = 1024) always left in the battery. That gives us 4096Wh of usable capacity. The 3 appliances I was comfortable powering with the starter module consume 743Wh leaving 3353Wh of reserve, roughly 4½ days' worth. That reserve calculation of 4½ days is based on zero solar charging. Unless you have 24 hours of night even a cloudy day gives some solar charging so your reserve will always be more than what's calculated.

With our 3-Day Blackout Power Plant, there was no way we could power all five priority appliances and I had to select three to use in our example. Now that we have more power available, let's again reexamine trying to power all five loads.

Our five appliances take 1201Wh. That is including the power used by the inverter. We have 4096Wh available so we can easily power all five loads if we had to with The Part-Time Power Plant Module. And you have almost 4 extra days of battery reserve in the event of a prolonged rain event. Don't forget, if you don't have a CPAP or a water pump to power, the power savings can be used for other things. I'm going through this exercise to show how the addition of the modules affects what can be powered and for how long while being consistent with our appliances in the different examples. That way I'm hoping it's easy to make some sense of all these examples.

For example, as an alternative, let's go back to the original three appliances, the refrigerator, smart phone and light. The original refrigerator in the 3-Day Blackout Power Plant was a small 5.8 ft³ model. With the addition of the One-Week Blackout Power Plant Module we increased the refrigerator size to 14.5 ft³ which consumed 923Wh of energy. Let's get an even bigger refrigerator. We'll go with a 21.3 ft³ version which consumes 1093Wh.

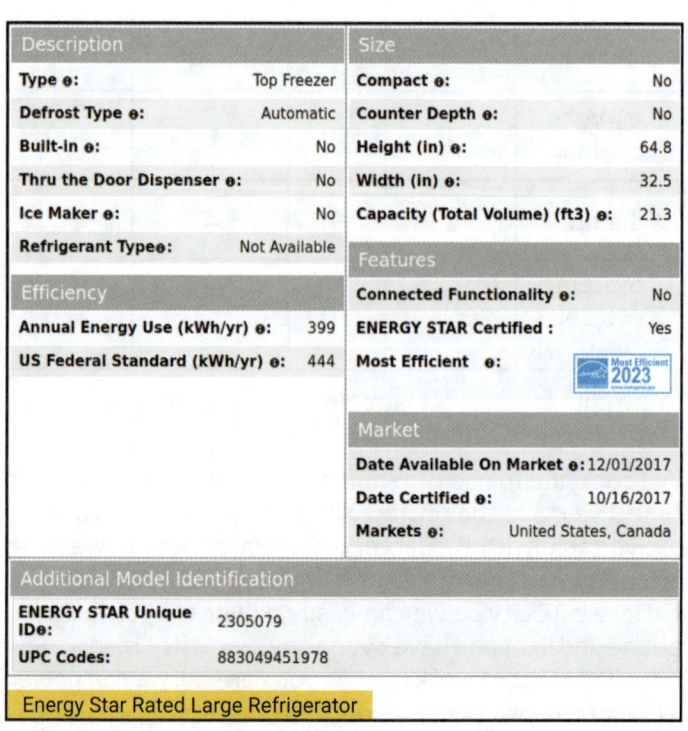

Energy Star Rated Large Refrigerator

Appliance/Device	Quantity	Watts	Hours Used	Daily Wh Used
Refrigerator (14.6 ft³)	1	136*	8*	1093
Smart Phone	1	12	1	12
Water Pump	1	700	.08 (5 minutes)	58
CPAP Machine	1	50	8	400
Lights	1	9	4	36
System Electronics Power Save		10	12	120

* I am guessing on wattage consumed and run time. But the 1093Wh is the needed number.

We continue to use the same table of appliances but this time, I've made the refrigerator more than triple the size of the small refrigerator we started out with. From the above table with the larger refrigerator added, we have an energy demand of 1141Wh. We have 4096Wh of usable battery capacity so we can easily power the refrigerator, phone and light and have close to 3½ days' worth of reserve left over. You can see a couple of extra batteries and solar panels makes a world of difference. As well, it sure does pay to keep energy efficiency in mind when shopping for appliances. If you go to any appliance site and seriously dig into the specifications, for the same size refrigerator, there is a wide range of energy needs to power the same sized device. It doesn't make much sense but that's the reality so shop with energy usage in mind.

Please keep in mind when evaluating all the numbers and scenarios I've laid out for you, my thinking is based on grid power down for the long term. Rainy days, cloudy weather and battery reserve will be of little concern if the power is only out a couple days. Unless you have confidence that power will only be out for a few days, then it's wise to simply plan for worst case and then get a nice surprise when the power eventually gets restored.

And finally, Johanna and I are firm believers in having backup plans, plan A, B and C. As much as we dislike generators, consider a small backup generator of 1000-1500 watts. It will be there as insurance. It likely won't be needed but if you've misjudged on power requirements, this provides an easy way to recharge the batteries if need be when a prolonged nasty weather pattern sets in. Only you know what your climate is like and can judge how likely that would be.

Although we have gone through the modules as I've designed them, you have the option if money is no object of buying up to 4 more batteries to wire in parallel with The Part-Time Power Plant Module. That's significant capacity at a significant cost. These particular lithium batteries work together up to a configuration of eight batteries total. So even though I have limited our three kits to a total of 4 batteries, it is possible to have a total of 8.

As well, regarding the solar panels, our charge controller is capable of dealing with 4 solar panels. There is no reason more panels and a second Midnite Kid solar charger can't be added to increase capacity. The only reason I stuck with the system I created is because this is not a permanent off-grid solution. This is purely temporary for when the grid goes down and you need to keep appliances running. If you do decide to add more panels, keep in mind you will need to store them in a location that protects them from damage. It would be a real drag to lose power and then find out a panel you counted on was damaged without you realizing it. So much for being prepared. That just went down the tubes.

Not only do solar panels need a proper storage space, but you have to lug each panel outside when needed. Not a big deal for some folks but it may be for others. Because I wanted to make this as reasonable as possible for the most number of people, I designed the modules with portability and storage when not in use in mind. If you do purchase extra panels, cost is not bad. Plus, there's nothing saying you must deploy all panels. How nice it would be to put out the 4 panels as designed, and if batteries start getting low, whip out another set of panels, especially if a couple days of no grid power turns into a week and then a month. You've now got plan A, B and C.

THE MODULAR BACKYARD POWER PLANT

CHAPTER 6

Basic Electric Measurement and Wiring

Typical Household Tools for the Job

Before we actually start to build our power module, this is a good place to talk about the tools of the trade, some basic techniques and more safety. Most households have access to basic tools. What you don't have can be borrowed. And there's nothing I'm going to list that can't be purchased at the local hardware or automotive parts store.

You will need the following:

- A black magic marker
- Tape measure
- Wrenches
- ¼" or ⅜" drive socket settings
- Electrical tape
- Solder and soldering iron
- A couple of screwdrivers both standard straight blade and Phillips
- Straight Edge
- Jig saw would be handy
- Crimper for crimp connectors
- Drill and bits/hole saw
- Diagonal wire cutter
- Rasp and flat file
- Long nosed pliers
- Sharp knife
- Insulated electrician's pliers or standard pliers
- Wire stripper
- Cheap digital multimeter

Basic Wiring

In order to construct our system, we need to know some basics on how to attach wires and how to arrange them in a neat orderly manner. You want something well-built you can be proud of.

As you already know, wires come in all kinds of sizes. We will be working with light to heavy wire. All of the cables and wire we use will have an insulation coating. Perhaps even a sheathing on top of insulated conductors (wires). Any insulation will need to be very carefully cut away to expose the metal conductor inside. Proper judgment is needed here. If you don't remove enough insulation, when you make a connection to a terminal, you run the risk of clamping down on the insulation instead of firmly clamping on the conductor which makes for a bad connection. If you take too much insulation off, you unnecessarily expose bare conductor which defeats the purpose of the insulation for protection. When it's time to wire, we'll go over this in detail.

Wire can be a solid conductor or it could be stranded meaning there are many thin conductors to form a larger conductor, the same idea as a piece of rope with lots of strands. When you strip insulation off a stranded conductor, before you do anything, twist that exposed wire to make it easier to deal with.

As well, when cutting insulation around the conductor, we want to be very careful we don't start nicking or digging into the conductor itself. Many times, on a larger cable you can score around the entire conductor with a knife and then with a little to-and-fro persuasion, break the insulation away without scoring down to the conductor. On smaller wire, we will use wire strippers which have varying size holes available for stripping different size wires.

As long as we know the wire gauge and use the right hole, it will safely remove insulation without damaging our conductor. How easy is that!

When we make any connection, ideally, we want not only an electrical connection but a mechanical one as well.

Proper Clockwise Loop for Screw Connection

As an example, if you've ever looked at a standard wall outlet installed in a home, the connections are generally made around screw terminals. I would strip the proper length of insulation and then with long-nosed pliers, I would create a clockwise loop from the exposed copper conductor. That loop would then be wrapped around the outlet screw making sure the loop is on the screw clockwise. I'd use the long nosed pliers to snug or crimp around the terminal to be sure the connection is secure and then I'd tighten the screw. Why are we specifying clockwise? When you tighten a normal screw, it is in a clockwise direction. As long as your loop was installed clockwise, it will have a tendency to tighten around the screw. If you installed the wire counterclockwise, it would try to uncurl your nice looped conductor as you tightened in a clockwise motion thereby creating a questionable connection.

Anytime we run a wire or cable into an electrical box or electronic component, we want to provide protection to that wire so that it doesn't get cut or chafed over time on sharp edges where the wire enters the box. We also want to provide some support to that wire. Any tugging on the wires going into an electrical box is really tugging on our connection itself. A strain relief will let us protect our internal connections so that the connection itself is never put under any tugging stress.

Typical Strain Reliefs and Backing Washers

THE MODULAR BACKYARD POWER PLANT

You may have seen some installations that run the cables in tubing. The tubing serves to protect the cables. We will not be using conduit (tubing) for our project and will only need strain reliefs in spots which is the standard way to wire electrical boxes and components.

When we run a wire from one point to another, we want that run to be neat and logical. We don't want our wiring job to look like a bird's nest when we're through. Not only does a neat job look good aesthetically but it makes any troubleshooting easier as you can follow the wires back and forth. We also do not want to run any wires to the exact length. We will run our wires slightly longer than necessary thus giving ourselves a little slack for easy maneuvering. If we ever needed to cut the connection off and make a new connection, we have enough wire to remove the insulation and make a new connection without having to run a brand-new wire replacement thanks to the extra length. That extra length of wire is called a service loop. When we are done our final wiring, we may opt to tidy things up with a few cable ties or even a couple of hold down clamps.

When we wish to connect wires together... well, there sure are a lot of ways to do it. Soldering, wire nuts, barrel connectors, terminal strips, low temperature solder connectors and push in connectors are all methods to connect a couple of wires together. Some of these methods are better than others and some are good for certain purposes. Although I may show a method I will use when I wire the system up, it doesn't mean another way won't work.

You are free to connect the wires in a different manner as long as the connection is strong electrically as well as mechanically. I understand we have many people building this who might not have access to a soldering iron or a set of crimpers so my choice is to give the easiest way the majority of my readers can make a good connection.

Not only must the connection be good, but we must provide that strain relief so that our connection doesn't degrade from pulling and general movement. And of course any connection we make must be well insulated so it doesn't short out against exposed metal or another nearby wire.

Multimeters

We've mentioned the multimeter a few times and it's time to know how to properly do a few things with it. Depending on how advanced the multimeter is, it can do a lot of different measurements. For the purpose of the backyard power plant, all we care about is doing some basic readings of voltage and continuity.

I can't know what model and brand your meter is so I'd suggest reading the manual that came with it as a starting point. You want to set your meter up to read volts. The following is generic info that would apply to any meter.

Your meter comes with a set of insulated leads, likely color-coded wires in red and black with a plastic handhold and conductor at the tip. Those test leads need to be plugged into the meter the proper way in order to measure a voltage. The black lead will be plugged into the "com" socket and the red lead into something that says "volts Ω" or something similar. The sockets may even be color coded themselves so you can plug red lead into the red socket and black to black.

Turn the meter on and set the meter to read DC volts. You may have a meter that automatically ranges or you might have a meter that you have to select a range. A range might be for example 0-10 VDC, the next range might be 10-100 VDC and another range might be 100-1000 VDC. These are just examples of what you might have with your meter.

Let's measure a small battery you have kicking around the house. If you can find an AAA, AA, C, D or even a 9volt battery, that will be great. The battery is a DC voltage. Our meter is set up to read a DC voltage. Because we are working with DC, we automatically know there's a positive and a negative. When you replace the clock or the smoke detector battery, I'm sure you pay attention to the polarity since installing the battery backwards will not work.

When you look at the battery, one end may be marked with a minus -sign and the other end with the plus +. If the battery is a 1.5 V AAA, AA, C or D cell, you now know what range to set the meter to. There's no point setting it to read 1000 volts. You ideally want to read the voltage as accurate as possible so make sure the meter is set for a lower voltage range such as 0-10 if that's an option your meter has.

Put the black lead on the battery negative (minus) side and put the red lead on the battery positive (plus) side. Your meter should come to life and if it's a good battery, you should read a positive voltage of something over 1.50 volts. Please repeat that a couple times to get comfortable with it. That's how easy it is to measure a voltage.

Measuring a Small DC Voltage

For fun, reverse your leads on the battery. It's OK. You won't damage anything but it may lead to some confusion. Your red lead is on the battery minus side and the black lead is on the positive battery post. See what the meter is reading? -1.50 You know the battery can't be wrong so when you reverse the meter polarity in relation to the battery, you may get a negative number. That's your clue to double check your meter leads since they are likely inadvertently reversed. When it comes time to measure our lithium battery, the procedure is exactly the same except we might set the meter range for 10-100 or whatever range is appropriate for your particular meter.

You've just taken your first DC measurement. Let's keep the momentum going. Let's measure the AC voltage of your house.

Measuring AC Outlet Voltage

This time, you don't have to worry about the polarity of your meter leads. This is alternating current that is constantly changing and the meter needs to be reset to read an AC voltage. Turn the knob and/or follow the manual's instructions for measuring an AC voltage. Carefully insert one of the test probes into the left slot of a wall outlet and the other probe into the right slot. **Make certain no fingers are touching the metal probes.** IMPORTANT: Your hands should only be touching the insulated plastic test probe handles. Your meter should be reading roughly 120VAC. **IMPORTANT:** Make sure you keep those probes apart so they do not touch each other or you will have a short circuit with blown circuit breaker as a result.

You have taken a DC and AC voltage measurement. The concepts for using the meter to read any kind of voltage will be the same. You need to keep the following questions in mind each time you wish to take a voltage measurement:

- Are my red and black test probe leads inserted into the proper sockets on my meter?
- Is it AC or DC that I want to measure?
- Do I have my meter set to read AC or DC?
- What is the voltage I expect to read and do I have the meter range set right?
- For a DC voltage measurement, polarity is very important. Black lead to negative and red lead to positive voltage.
- For a typical AC voltage measurement, we do not need to be concerned with polarity.
- I want to be very careful when I measure voltages of any kind so that my fingers and hands are only grasping the insulated test probe handles and no fingers are touching the metal test probe tips.
- I want to be very careful when measuring so that I do not accidentally touch the probes together while taking a measurement.

Regarding that last point, there are times such as measuring the voltage at a wall outlet where the two points that need to be measured are very close to one another. Care must be taken that the two points do not get short circuited by the meter test probes.

The last thing we will do is measure resistance/conductance. You have your meter's manual to refer to if anybody ever asked you to measure resistance accurately. While we will set the meter to read resistance what we really are asking the meter to tell us is if we have an open wire or perhaps a short circuit.

The first thing to do is **TURN OFF THE POWER** to anything we wish to measure with our meter set for resistance. Unlike reading a voltage where our circuit or voltage source is live, when we go to measure resistance no voltage can be present. To be absolutely safe and certain no voltage exists, disconnect the positive terminal of the battery so there's no chance of battery voltage in the circuits. As well, disconnect the positive lead of the solar panels. The easiest way to do this is to use the plastic tool made for disconnecting the connector out at the solar panel. We've eliminated any chance of power coming into our system. Now we can safely check for continuity or resistance.

As a simulation, just grab a short piece of wire. Make sure exposed conductor is showing at each end. Set your meter to measure resistance or if you have a beeper setting, put the meter to beeper. Put your meter lead on one end of the wire and the other lead to the opposite end. It doesn't matter which lead goes where. In this case, we don't care about polarity.

Your meter should read close to zero when you make the connection and/or the beeper sounds signifying you have a complete circuit. Your wire is good. Now, take one lead off and hold it into the air while still keeping a lead on the one end of the wire. The beeper will stop beeping and the meter has no reading. You have simulated a break in the wire somewhere. In other words, the bridge is out. You now have an open circuit. The current doesn't have a path to travel. Like a road with a bridge, if the bridge is out, the car doesn't get across. Without the intact wire, the current can't flow.

That's the general concept. But I should also point out while we are here, there are times when some daredevil sees a bridge out and decides he/she is going to get a head of steam and jump it. Electricity is funny stuff and under the right conditions, it will also jump the open bridge. We call that an arc or spark. In an extreme case, we call it lightning.

At this point, you might be scratching your head and muttering out loud "Big Deal – So what does this all prove?" How is this helpful? You can't imagine how useful that little bit of meter knowledge can be. The first time you find something doesn't work, yet it sure looks connected and you're left wondering what's wrong; all you need to do is pull out your trusty meter to confirm there really is a connection. Then lo and behold, your meter says the wire seems to be broken somewhere. So you decide to unscrew one end and find that the wire was clamped on to the insulation and wasn't making contact. Or you find the screw holding the wire down was never properly tightened even though it looked tight visually.

Remember, the resistance of the wire should be very close to zero. If you measure a wire expecting to read zero and there's significant resistance, pull the wire off and you likely will find it is corroded and not making a good contact connection. Your meter just became invaluable for troubleshooting.

I could rattle off all kinds of scenarios where continuity checks would be so helpful but you get the idea. If ever in doubt that a wire really goes where you think it goes or a wire doesn't seem to be making contact or you are wondering whether the chassis of a device is really grounded to negative, your meter will be able to cipher it out for you and confirm it.

The thing to remember is whenever you want to try reading resistance, the circuit or object must not have any power to it. We will do more with meters later.

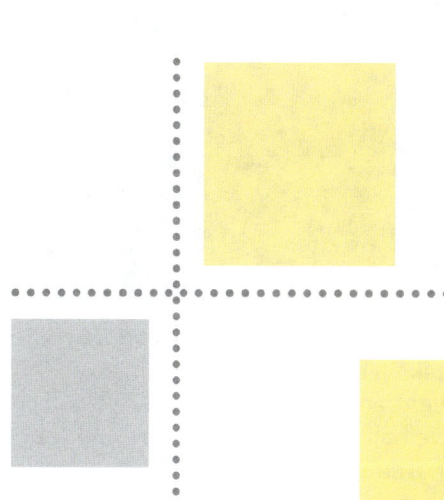

THE MODULAR BACKYARD POWER PLANT

CHAPTER 7
Let's Build Our System

3-Day Blackout Power Plant Assembly

We've spent considerable time going over basic electricity, theory of designing a system as well as how each of the components work together and it's now time to put everything together and make it play.

We'll start with our solar panels. There are fancy mounts we could have bought for our temporary situation and of course, various mounts for permanent installations. They cost money and I am into saving money if at all possible. In fact, at the time of this writing, the mount suggested for each of our panels costs $79 a piece. So, I gave some thought as to how I could easily set up one or more panels on a lawn without breaking the bank. As a result, we can build mounts for all 4 solar panels for around $30 total. I can live with that. Here's what I came up with.

We will use the following materials which I bought at my local building supply/hardware store.

This list of materials will work for one solar panel:

- 1 piece of ½" X 10' rigid PVC plastic electrical conduit
- 2 pieces of ½" PVC 90° elbows
- 2 pieces of ¾" EMT Steel single hole straps
- 2 pieces ³⁄₁₆" X 1¾" Linch pins
- 2 #10 X ½" screws, flat washers and wingnuts for quick assembly

The following directions will be for a single panel so double or quadruple the materials if you have two or 4 panels total respectively.

Note our conduit has a flared end so another piece of conduit can be added. We want to work starting from the opposite non-flared end.

Flared Conduit End

With a tape measure, measure off 24 inches from the end and make a mark with a magic marker. Instead of cutting it straight across at a 90° angle, we will cut at a diagonal 45°. Eye ball is good enough here. Do your best to make a diagonal cut with a hacksaw.

45° Diagonal Cut Using Hacksaw

Now use that piece as a template to cut the next piece by holding it next to the long conduit such that the angled ends are lined up at one end.

First Cut Piece Used as Template

Now make another mark at the other end. This time we will cut straight across at a 90° angle. We now have two pieces that are the same size and configuration which will be our legs. Each piece will have an end with a straight cut while the opposite end has a cut at a 45°. The ends with the 45° angle will make it easier to push into the earth and will help to anchor the panel to the ground when it's time to set it outdoors.

Take the 90° elbows and push them firmly on to the squared cut ends of the two legs. They should be tight. Make sure they are firmly seated.

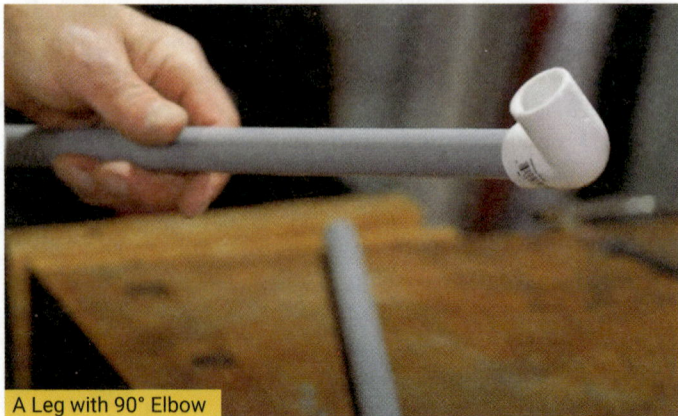
A Leg with 90° Elbow

I would prefer not gluing our pieces together. For one thing, the pieces are made to fit snug and will work fine in this manner. As well, being able to pull everything apart gives you versatility. Who knows if/when you may need the pieces for another project or emergency and the piping and fittings would prove invaluable to you.

Try to have a mindset that gives you as much flexibility as possible. These items and parts can be reused for other things depending on the circumstance. There may be a time when the situation is more serious than a couple days without power. A piece of tubing might just be the thing to get something running and have a better use than holding up a solar panel. You just have more versatility if you have the option to pull pieces apart later for another purpose. Although the pieces should be snug, with a pair of pliers, you can easily wriggle everything apart if need be.

Getting back to constructing our solar panel frames, from the remaining length of tubing, mark off and cut 2 pieces that are 3" long. Insert each 3" piece in the other side of the elbow. Again, fully seated and snug. Set those two legs aside for a moment.

Looking at the back of your solar panel, you will see multiple drilled holes already available for mounting our panels. Our panels when set out on the lawn will be low profile, meaning they will sit on the ground long side down. In picture parlance, oriented in "landscape". With our solar panel resting on the ground long side down, find the top hole on each side.

We will use those holes to mount our legs by using an EMT metal strap fastened with a screw and a flat washer which comes in from below, through the hole. Then the metal strap goes over the screw and a wingnut goes on to the screw holding everything in place. Orient the strap up towards the sky and then tighten the wingnut.

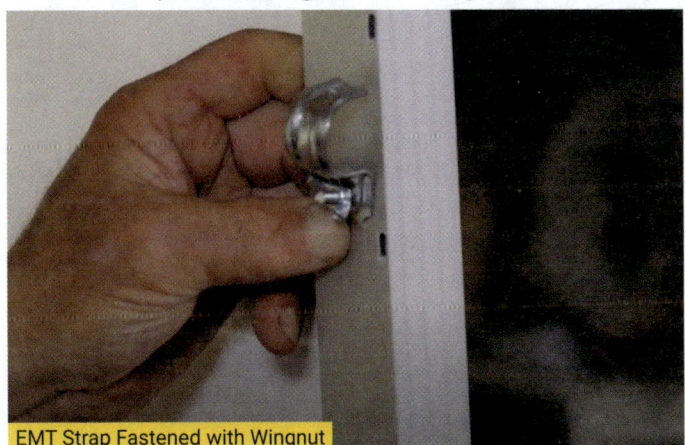
EMT Strap Fastened with Wingnut

Repeat on the other side. When completed, our solar panel should have a fastened steel strap on each side of the panel.

We can now set our legs in place. Slide a short leg through the strap from the inside of the panel facing outwards. You should have about ¾" of tubing extended beyond the sides of the panel. With a marker, make a mark on that short stub about ½" from the end.

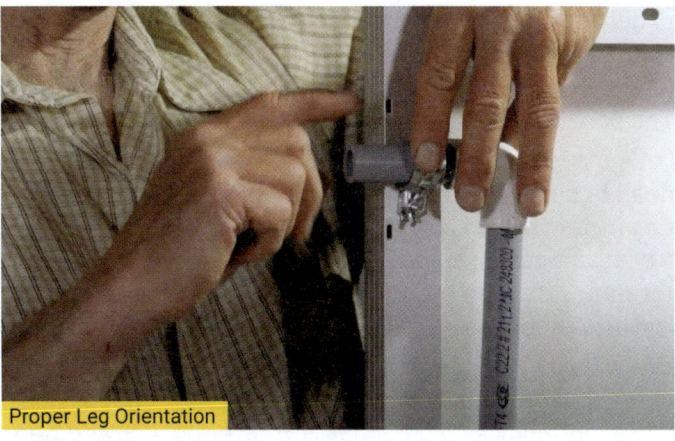
Proper Leg Orientation

The idea is to drill a ³⁄₁₆" through hole with a drill to insert our linch pin which will hold the assembly together. Make sure your mark will allow enough room to do that.

Take the legs off and drill the hole straight through your short piece of tube at the mark you made on each leg. Take your time and try to make sure you are drilling straight down through the tube. It would be best to lay that tube on a piece of scrap wood so when your drill exits the far side, you won't damage anything.

Through Hole with Drill Oriented Vertically

Reinsert the legs into the straps and put in your linch pins. Your solar panel legs should pivot easily in the strap but can't be removed without pulling the linch pin.

Leg Installed with Linch Pin

The solar panel is now ready to be stowed ready for use. For storage, it might be easiest to pull the linch pins, pull the legs off, put the pins back into their respective holes so they aren't lost and then store panel and legs in a closet or basement until needed.

When the panels need to be deployed, I found the easiest way to set them up is to pull the legs back in unison and push the legs into the soil a few inches. Then adjust the panel angle by pulling the base of the solar panel forward to the appropriate angle. Then, if need be, tweak the adjustment by pushing the legs a little deeper into the soil. The panels should be well anchored and stable.

Let's move on to our lithium battery. The first thing we need to be aware of is there's a lot of potential energy sitting in a small package. We don't need to be afraid of it but a dose of healthy respect and an equal dose of thinking in terms of safety will go a long way to making sure we don't run into problems.

We want to don a pair of safety glasses anytime we are working around batteries. This is always paramount when I do maintenance on our lead acid batteries such as when I have caps open and am busy measuring specific gravity of each cell. An errant splash on a finger or skin is one thing as this can be flushed immediately with water with no ill effects if done quickly, but a splash in the eyes is a game changer. Why take a chance? And why not carry that safety regimen through to any battery regardless of type? We may have some inexperienced people putting this system together and wearing safety glasses/goggles is a good habit to get into.

Another thing we must realize is the battery by its very nature is a source of energy just waiting to be used. It will happily supply current to any load that shows up between the positive and negative terminals. And that includes a careless drop of a wrench or wire that makes the connection across the battery terminals. That would be catastrophic. I can't be any blunter. Never short circuit the terminals of any battery.

By short circuit, I mean essentially having a conductor such as a wire, wrench, screw driver, necklace or piece of metal somehow make contact across the plus and minus of the battery. You need to be aware of how you store the battery when it comes time to put it away until needed. Putting your solar panel with its nice aluminum frame next to the battery is a bad idea for example. Ideally the battery would be set aside and stored safely in the basement on a shelf or other safe area or perhaps in the garage. The battery came with a couple of plastic protective caps covering the posts. Put the two covers back on those battery posts when the battery is in storage for your primary protection.

I'm not only interested in making sure your battery is safe from accidentally having a stray piece of metal make contact with the terminals, but you want to have easy access to your battery for maintenance. The only maintenance your lithium battery needs is the occasional charge to keep it fully charged. Remember, over time, your batteries will have a tendency to slowly discharge. Your batteries should self-discharge less than 3%/month. Over 3 months, you might be down close to 10% discharged on your batteries. As a matter of good practice, every 3-4 months, make sure they are fully charged up. You will be behind the eight ball if you lose power and have to run your appliances starting with anything less than a fully charged battery.

We can easily do this with your battery charger which runs off house power. There's no need to set up the solar panels and go through the process of charging batteries as long as you have grid power and the charger I specified. Although the loss of grid power can come at any time, if you have advance warning that weather or a brown/black out is bearing down on you, that's the time to make certain the battery is fully charged ready to go before the grid goes down.

Since I have never used lithium batteries, this will be a learning experience for me as well as you. We will follow the easy steps outlined by the battery company. Let's take out our voltmeter and set it on the volts. Take note on our battery there are two terminals. A positive (+) and a negative (-). From this point forward, we **MUST** know which terminal is positive. Although it is well marked with a plus sign and spelled out, feel free to put a piece of masking tape on that end of the battery with the word "positive" clearly denoting which terminal is which. Note as well, each terminal is color coded with the positive terminal having a red band and the negative terminal a black band around the post.

Take the black lead of the multimeter and put it on the negative battery post. Then take the red lead and put it on the positive battery post. What is the voltage?

Measuring Battery Voltage with Multimeter

Don't be alarmed if it reads zero or close to it. That just means you need to wake your battery up. We'll do that by applying a charge with your charger. We'll follow the directions that come with our charger.

The charger I purchased for use with your batteries is made specifically for 24V lithium batteries. Check with the battery manufacturer on charging specs if you want to use a different charger. The manufacturer of the battery I've specified recommends a special charger with preset charge parameters especially for lithium. Since the batteries are expensive and this is for emergencies, I wasn't willing to gamble and take a chance on overcharging and damaging the batteries by purchasing any other charger than what the manufacturer recommended. Hence, my purchase and suggested purchase of this special charger for you to use with your lithium batteries.

In theory, you could charge each battery individually assuming you started out with more than a single battery and then you know each battery is fully charged. But you should be able to charge all connected batteries as a single large battery which would be no different than how your solar panels and charge controller would work.

This particular charger I specified is nice in that you can keep half the connector attached to your battery bank safely and if you need to charge them, just make the connection to the charger with the mating part. So at this stage, let's get the batteries charged, activated and ready for use when the time comes.

If you've ever charged a car battery, the same principle applies. I had to slightly modify my charge cable to span the width of the battery posts and you probably will have to do that to. All I had to do is carefully cut the insulation back. Peel the shrink tube back and remove enough insulation the wires can be connected to each post. Then put the shrink tube back on.

Modified Battery Cable

As you know, the battery has a positive and negative post. Very important you know the difference. The charger connector has a red and black wire. The red lead must go to positive and the black lead to negative.

Let's put on our safety glasses. This is quite safe but I'd like you to get into the habit to do this every time you work around a battery. Unscrew the two terminal posts. Connect the red and black wires. Red to positive and black to negative. Tighten the bolts snug. Don't over tighten.

Battery Charger Cable Connected

That connector is just flapping in the breeze now. Plug the charger in. Take note of the green LED lights on the end.

THE MODULAR BACKYARD POWER PLANT

Green LED's visible

That's good. The charger is alive. Now let's plug the male connector into the female and snap it into place to make the connection. There's only one way the connectors will go together so don't worry. You can't go wrong here.

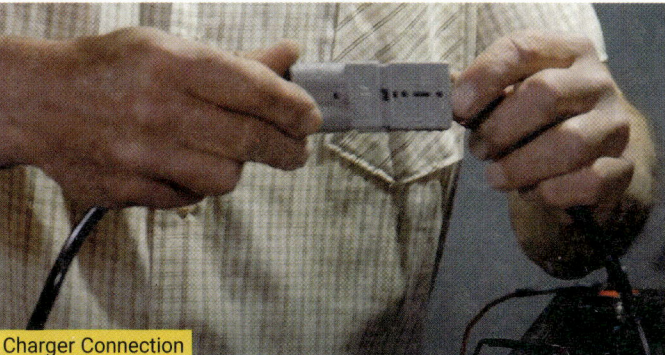
Charger Connection

Note the LED lights at the end just turned red.

Battery Charging

That too is a good sign. The battery is charging. Just let it remain in a safe place charging and wait until the green light returns. The charger will turn off and your battery is fully charged. Do that for each lithium battery you have. You can pull the connector apart to disconnect. It will take some effort. Give it a good pull but don't pull by the wires. See, this isn't so bad if we just take small chunks at a time. The batteries are ready to go and it's time to build our power plant.

Mounting All Our Components

You have several boxes of magic electronics that have been purchased. Everything needs to be mounted in a fashion that is logical, orderly, that allows you to view any monitoring lights and displays and that lets you make the shortest, neatest wired connections possible while at the

THE MODULAR BACKYARD POWER PLANT

same time allowing you to plug your gadgets in when needed. You want it to be as portable as possible and you need to be able to stow it in a closet or area out of the way until needed.

When I laid this all out, I knew I could have made this more compact. And there's nothing stopping you from doing that if you wanted. I chose this size for many reasons. Ease of looking at the display, ease of plugging devices in both on the inverter itself and the USB power module and I knew this would ultimately be filmed and I wanted things nicely spaced and easily visible.

Refer to the following illustration for the dimensions of the components on a mounting board.

to make the "feet" for our sheet so that it can free stand with all of our mounted components. It is important we make sure the feet are adequate and have a wide enough stance so that our power center doesn't topple over. That would be very bad. The part of the bracket sticking out on the floor acting as a support should be at least 7 inches long.

It would be helpful if you had someone available to hold that panel vertically on a table top as you set each shelf bracket in place and screwed it in with those ½" wood screws. Repeat until you have all 4 brackets fastened and your board is perfectly vertical and freestanding on the table top.

Power Plant Dimensions

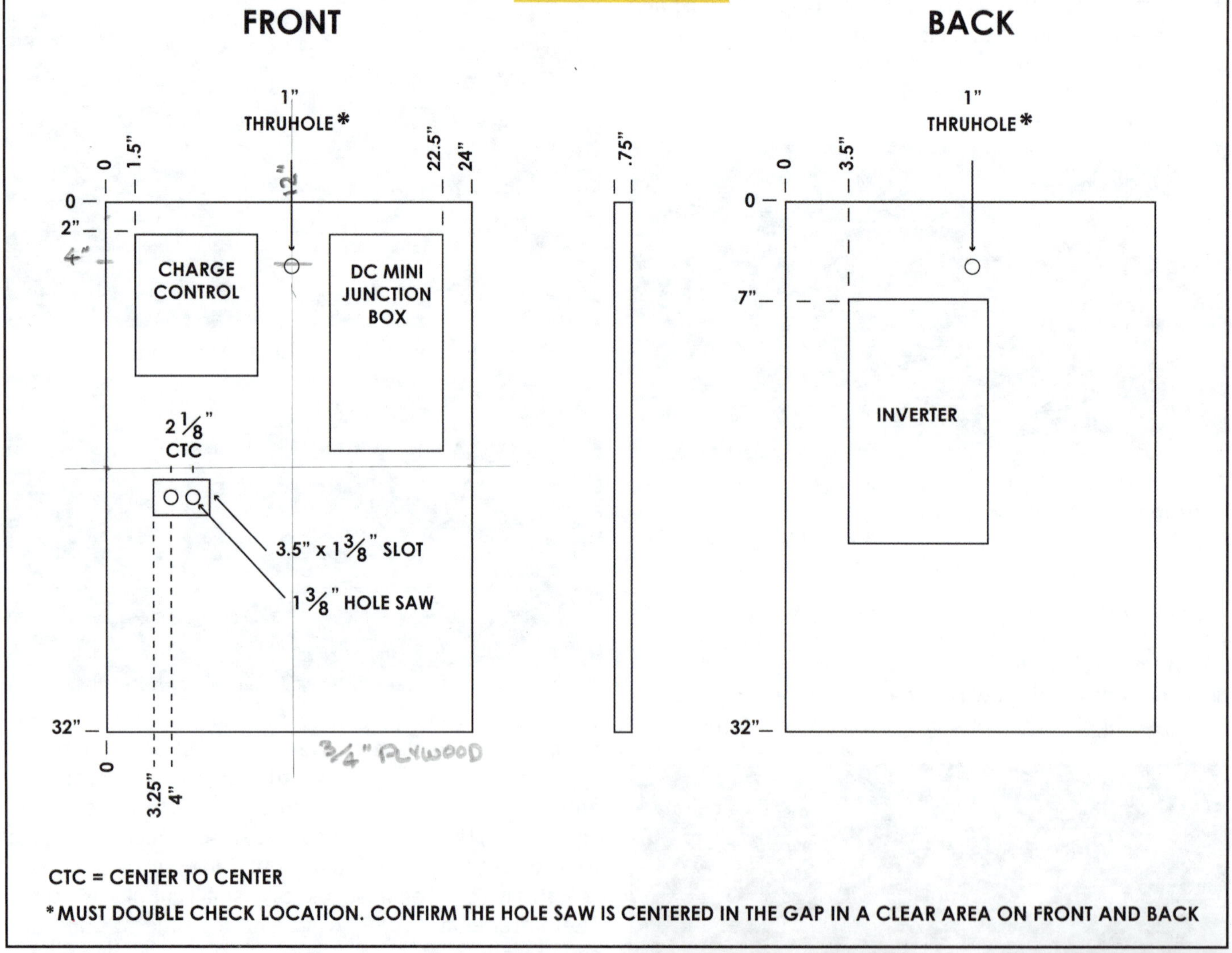

[Diagram showing FRONT and BACK views of the power plant board with dimensions. FRONT: 24" wide x 32" high, with 1" thruhole at 12" from top, Charge Control box at 1.5"–? and 2"–? from top-left, DC Mini Junction Box at 22.5", 3.5" x 1 ⅜" slot and 1 ⅜" hole saw with 2 ⅛" CTC, measurements 3.25"/4" from left at bottom. BACK: shows 1" thruhole at 3.5", Inverter box at 7" from top, 32" high. ¾" PLYWOOD noted. Side view shows .75" thickness.]

CTC = CENTER TO CENTER

*** MUST DOUBLE CHECK LOCATION. CONFIRM THE HOLE SAW IS CENTERED IN THE GAP IN A CLEAR AREA ON FRONT AND BACK**

You'll need a piece of ~~plywood~~ or particle board with a ~~thickness of ¾"~~. Many of you probably have a partial sheet of plywood or OSB kicking around that can be used for this. You will use ½" ~~#10 wood~~ or sheet metal screws to mount all your components on both front and back of this sheet so this gives you plenty of thickness without worrying about your screw protruding through to the other side where it shouldn't have.

~~Cut the sheet to 24" wide X 32"~~ high. In the interest of saving money, I used some spare shelve brackets I had laying around

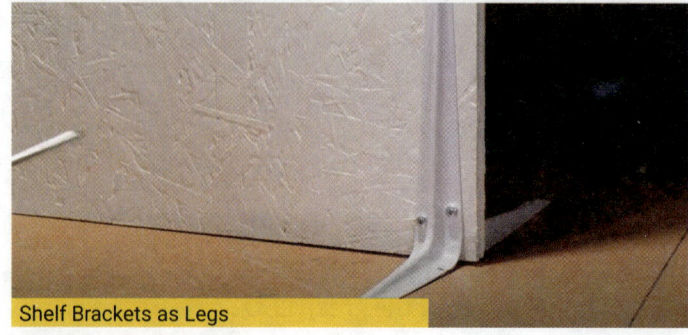

Shelf Brackets as Legs

Shelf Brackets as Legs, Another View

If like me, you want to get going without waiting for another pair of hands, you can do this by yourself by using a wooden wood clamp. Placing one on an end will suffice to lock the board vertically. Do a good job here. This is not the place to create the leaning tower of pizza (That's how you spell Pisa when you're hungry).

Lay the sheet flat on a table and from the top, mark down 2" from the top in a couple places. With a straightedge, connect the marks so that you have a line going across the sheet 2" from the top.

From the left side, mark off 1½" from the edge in a couple places down the side. Again, with the straight edge, connect the marks. Repeat the process on the right-hand side of the sheet making your marks 1½" in from the edge. Your solar charge controller will fit within the box on the top left and your mini-DC junction box will fit within the top right of our marks.

Turn the sheet over and make a couple marks 7" from the top. Connect those lines.

Using Straightedge to Mark Locations

From the left-hand side, make a couple of marks 3½" and connect those lines. Your inverter will fit into that marked box.

Turn the sheet back over to the front.

Charge Controller Oriented Properly

Orient the charge controller so the display is to the top and the wire entry ports are at the bottom of your sheet. Use four ½" screws to fasten the charge controller to the sheet, two screws per side.

Moving on to our mini-DC enclosure, remove the screw holding the lid and slide the lid up and off. Inside will be some hardware and a large 175 Amp circuit breaker. From here on out, we will call that big breaker the main inverter breaker.

Now take a close look at your mini-DC metal enclosure again. Notice there are many stamped potential openings, properly known as knockouts, which can be used to feed wires in to be connected.

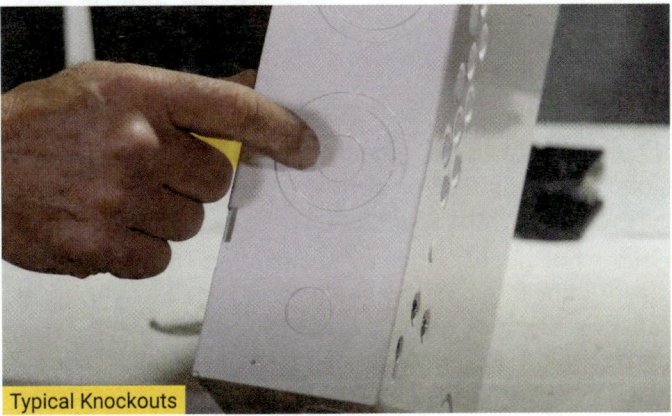
Typical Knockouts

It's similar to any electrical breaker box knockout except there are different sized holes that can be punched out to accept a strain relief connector. At the bottom of the mini-DC enclosure, there are multiple knockout spots. You will want to create a hole at the bottom left. That position has three different sizes that can be made depending on how much metal you remove. In order to remove the inner 2 rings, take a hammer and flat screw driver and punch those 2 inner rings out. Once you get it started in a place you can pry it loose with the screwdriver and then use pliers to wiggle it free. Try to leave that 3rd outer ring in place. If by chance it breaks out with the rest, you will need to go to an electrical supply place and get a couple of larger flat washers to go with your strain relief connector.

Insert and tighten your strain relief. Use a hammer and flat head screw driver to tighten the nut on the backside.

On the left side of the mini-DC enclosure (near the charge controller) you will install another large strain relief in the 3rd knockout from the bottom. Again, try to leave the third outer ring and only remove the two inner knockout pieces. Once installed, you can again use a hammer and screw driver to tap and snug the nut.

The mini DC junction box should be oriented so that the circuit breakers are facing the edge of the sheet towards the right for easy access. With the box properly oriented, use four ½" screws to fasten the box to your sheet. Line up the top edge with the same mark you used for the charge controller and line the right side up with the line you made previously on the right side. Make sure it's aligned neatly and put two screws in at the bottom holes. Look good?

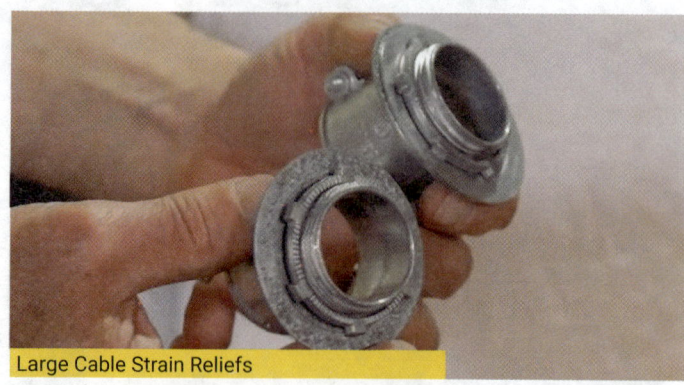
Large Cable Strain Reliefs

Let's take a minute to install the three circuit breakers. From the assortment of mounting screws in the package, take out the 2 machine screws that will be used to mount the main inverter breaker. Those machine screws are slightly different from the rest of the screws in the package. There are only two of that kind. The top slot on the narrow side of your mini-DC box will be the spot for the big breaker. But it probably won't fit. Try to fit it in the hole. It should seat completely through the hole. Take note that there is a removable piece of metal on each side of that hole that once removed, will allow that breaker to sit properly. The reason those removable side tabs exists is to accommodate different sized breakers. With a pair of pliers, grab on to those tabs and wiggle back and forth until they break off.

Hold the breaker from the inside and fasten the breaker using the 2 machine screws from the outside. Those two screws will come in from the front, go through the metal side and direct into the breaker itself.

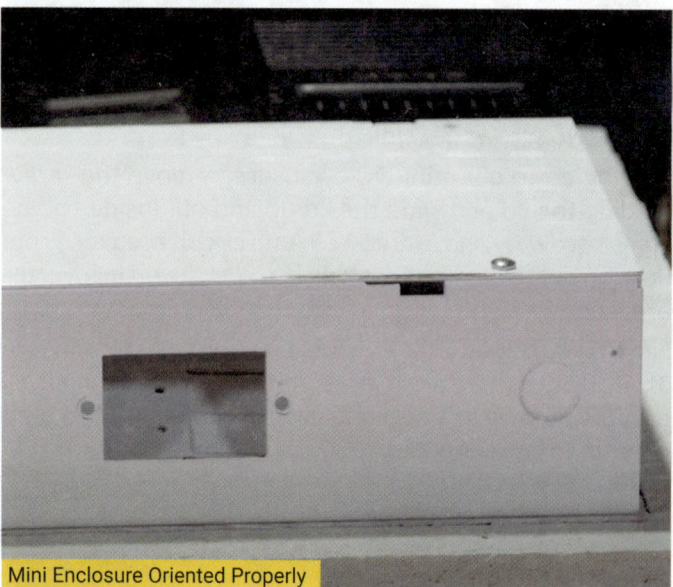
Mini Enclosure Oriented Properly

Main Inverter Breaker Mounted

If so, then put the remaining two screws in the keyhole slots at the top making sure the head of the screw captures the top of the keyhole slots.

On the bottom of our charge controller, The Kid, are two plastic knockouts. With a hammer and screwdriver, gently tap those 2 knockouts and break them free. Install 2 standard strain reliefs made for the typical electrical junction box. They will look like this.

You want the breaker oriented so that when you look at your enclosure from a front view, when the breaker lever is down, the breaker is OFF. When you flip the breaker lever up, you have now energized your circuit by turning the breaker ON. Essentially the same idea as your light switch in your house.

You also ordered a couple extra smaller circuit breakers and you can mount those as well now.

THE MODULAR BACKYARD POWER PLANT

Two Smaller Circuit Breakers

These breakers snap into place. There's a rail near the opening below your main breaker. By sliding the breaker and catching the bottom lip, you can now pivot the breaker back so it snaps into place. Before installing, make sure you orient each breaker so that they too will be in the off position when the lever is pointing down.

The slot below the big breaker has a metal bracket nearby. Look at the bracket and note there are two protruding channels.

Circuit Breaker Mounting Bracket

Your smaller breakers will mount on that bracket. Take a look at the 20A breaker. Orient the breaker with the yellow tab to the bottom. When you observe the back of the breaker, you can see an indentation channel. And that yellow tab will slide down to open another channel. Slide the yellow tab down.

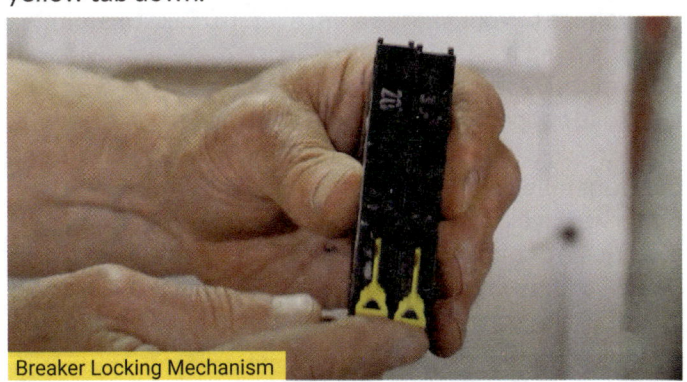
Breaker Locking Mechanism

Then insert the small breaker on to the bracket below the big breaker hooking the top breaker indentation on to the bracket top channel. The breaker should sit on there cleanly. Push the bottom in and then slide the yellow tab up to lock the breaker in place.

Circuit Breaker Mounted

You should not be able to get that breaker off unless you slip that yellow locking tab down to release it. However, you should be able to slide it to-and-fro on the bracket. Slide it back roughly 1½" from the edge of the slot. Repeat the exact same procedure with the 30A breaker and slide that roughly ½' from the front edge of your slot. Your three breakers are ready to wire in.

Before you mount your inverter, let's take a moment to configure it properly. Remember back in our design chapter, we talked specifically about our inverter and the amount of power it draws when our unit is sitting in idle, idle meaning there's no loads on at the moment. And yet, our inverter is taking enormous amounts of power sitting there waiting for something to come on. That's hardly an efficient utilization of our batteries. However, our inverter has some settings we can modify to go into "energy saving" mode. Not great but definitely very helpful in saving our battery power for things we actually want to run.

Take a look at the top of the inverter where the outlets, switch, LED display, connectors etc. are located. There are 8 DIP (Dual Inline Package) switches.

THE MODULAR BACKYARD POWER PLANT

DIP Switches to Configure

They are tiny little things. If you have a set of mini screwdrivers or a toothpick, you will need them to make some changes to those switches. Be gentle. It doesn't take much to move them and they only slide a teeny tiny bit in each direction.

You will leave switches S1, S2, S3 and S8 alone. Just confirm that all 4 switches are "ON". With S1, S2 and S3 set to on, your voltage will be 120VAC@60Hz which is just what we want. S8 being on tells the unit to look at the DIP switch settings. Switches S4, S5, S6 and S7 will be off when the unit comes to you and it's a couple of those switches we will change.

Leave switches S6 and S7 in the "OFF" position. Slide switches S4 and S5 to the ON position. You have now set your inverter up for the most efficient operation it is capable of. A 4% threshold power saving level.

Your unit is rated at 2500 watts. The threshold is 4%. Your unit should go into power saving mode if your total load drops below 100 watts (.04 X 2500) for 10 seconds. Any load above 100 watts should reawaken the unit. You will wire in and run your light and USB charger direct off your battery to further capitalize on the setup.

On the reverse side of your board, you need to mount the inverter. The easiest thing is to lay the sheet down on a table with the front side and components protected by a blanket. Set the inverter down on the sheet lining the left and top edges of the inverter to their respective marked lines you made. Use four ½" screws, two per side to fasten it. Since the mounting holes are so large, I used a flat washer on each screw to make sure I captured the entire mounting slot.

Fastening the Inverter with Screws and Washers

Wiring Our System

Now that you have all your components mounted in the right locations, it's time to wire this up. You have 12 wires to connect and we'll knock them out one at a time. First stand the unit back up on the table top with the charge controller and mini-DC enclosure facing you.

Find your 5' red and black #1 AWG battery cables. Those cables have a ring lug to make connections a cinch. Take the black cable and feed it through the strain relief you just installed on the bottom. Slide the ring lug onto the welded threaded post. Find the washer and nut and loosely put it on to keep that cable in place.

Installed Negative Battery Cable

You will need a short jumper cable to supply battery positive voltage to your charge controller. To make this jumper, you will need a 12" piece of #10 wire preferably with red insulation. Depending on your wire source, this wire could be a solid copper wire or it could be stranded. One end will need to have a ⅜" hole ring lug crimped and/or soldered on. Hopefully your solar supplier made this piece for you. Otherwise, you can make it or an electrical supply house, electrician or handyman buddy should be able to whip this piece up in a minute for you. Strip ½" insulation, insert it into the ring connector and with crimpers, squeeze that connector tight. If this was a stranded wire, twist the wire

THE MODULAR BACKYARD POWER PLANT

tight before inserting and crimping. Give the wire a tug to confirm you did a good job. On the other end, strip off ½" of insulation by carefully scoring the perimeter with a sharp knife enough to break it free.

Find the bottom post of the 175A main inverter breaker. Take the nut, lock washer and flat washer off. Feed the red battery cable through the same strain relief as the black battery cable. Take the red cable and attach it to the bottom post of your 175A main inverter breaker by sliding the ring lug onto the post.

Installed Positive Battery Cable

Slide the ring lug end of your short jumper wire on to this post, then put the flat washer, lock washer and nut back on in that order. You should have two cables attached to that post. The red positive battery and a jumper cable. Tighten securely. Your battery cables are now done.

On the front of your mounting sheet, just between the charge controller and the Mini DC enclosure is a narrow gap of 2 inches. Measure off 3½" from the top down. You need to bore a 1" through hole to feed a couple wires to the inverter. The easiest way is to use a drill and a 1" hole saw. If you don't have a hole saw, a hole needs to be made one way or another. Make absolutely certain you aren't drilling or cutting into your inverter. Double and triple check your measurements.

Find the two cables both red and black that are 3' long. There will be ring lugs at each end. Slip the ends of both cables through the side strain relief.

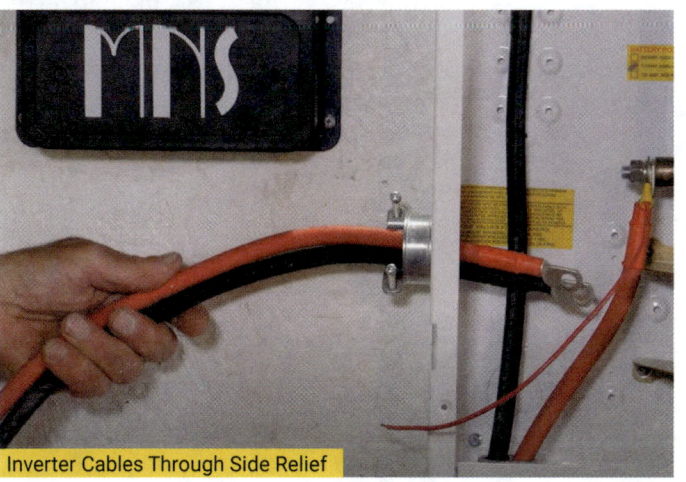
Inverter Cables Through Side Relief

Unscrew the nut you loosely put on the post with the already attached black battery cable. Attach the black 3' inverter cable, put the nut back on and tighten snug. If you have a torque wrench, torque to 180-inch pounds.

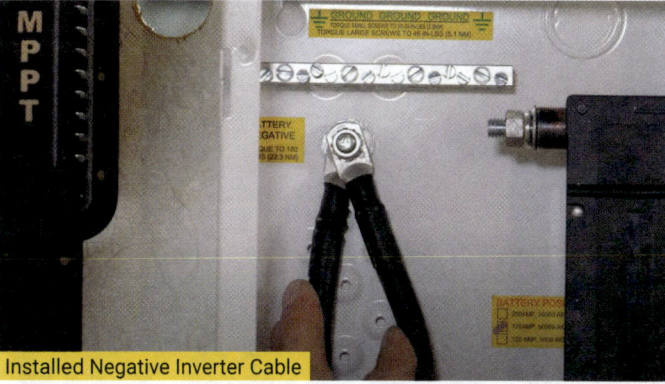
Installed Negative Inverter Cable

The red cable will go to the top post of your 175A main inverter breaker. Take the nut and washers off, slip the red cable ring lug on to the post and then put the flat washer, lock washer and nut back on in that order. Tighten securely.

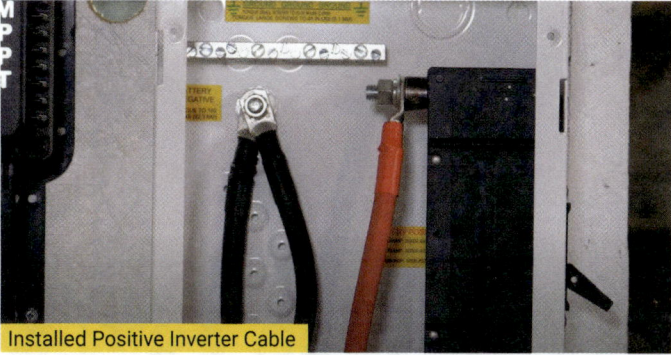
Installed Positive Inverter Cable

Let's turn our attention to those two dangling cables coming out of the mini-DC box. Those are your power cables that go to your inverter. Feed both wires through the 1-inch hole you recently made. Notice how the wires neatly align with the inverter input? All you need to do is take out the two screws where the cables attach. Slide the ring connector in and pop the screw through the ring and tighten snug. Note that the ring lug is offset from the cable.

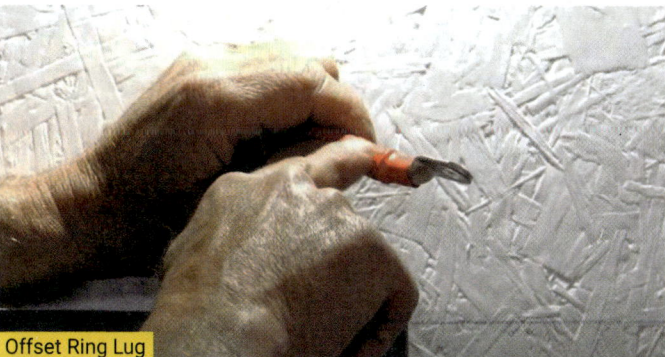
Offset Ring Lug

There's only one good way these lugs will fit into the opening. In one direction, the wrong direction, the lug will hit against the plastic housing. If you reverse the lug, there's plenty of clearance. It is imperative the red cable go on the battery positive side and battery negative go to the negative input side.

45

You can easily tell which is positive and negative. If you look closely, there are markings "Pos +" and "Neg -". You will also see a red and black band between the two connection points.

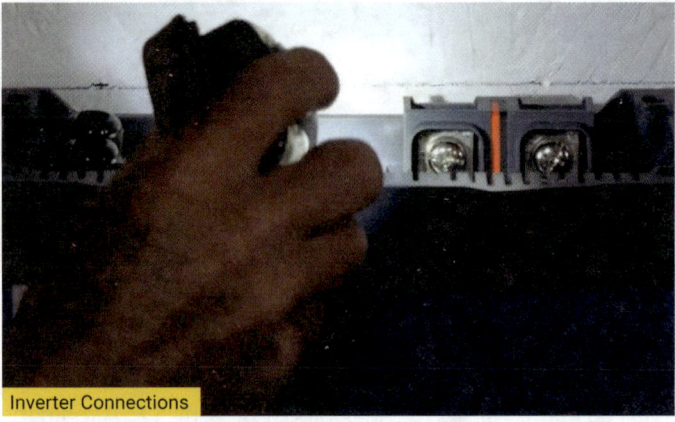
Inverter Connections

Your red cable is to go on the side with the red band and the black cable goes to the black band side. Use those as a double check that you have them right.

Fastening the Inverter Cables

Let's connect the dangling red insulated 12" jumper that you attached to the bottom post of the 175A main inverter breaker. Carefully and neatly bend the wire down and around the bottom end of the 30A breaker. Insert the stripped end of the conductor into the opening and by using a Phillips screwdriver, tighten the wire snug. Test the connection by giving a tug. The wire clamped well?

If you have trouble making this connection, just reverse the order. You can disconnect the red jumper from the bottom post of the 175 main inverter breaker, take out the 30A breaker and while it is your hand, simply make the connection to the bottom terminal, reinstall breaker and then reconnect the ring lug. That works fine as well.

Take the 2 screws out that are holding the charge controller cover in place so you can access the wiring. Note there is a terminal strip with one side wired already. The other side is where you will make your connections. The slots are marked with what wire should go where. Refer to the manual as well.

Take a 24" #10 black insulated wire and strip ½" of insulation off one end and ?" insulation off the other end. Feed the end with the ½" of insulation removed through the side strain relief. Neatly feed and bend it up so that the

conductor can be inserted into one of the many holes on the ground bus bar. Choose one of the smaller holes appropriate for the diameter of your wire. Tighten snug and give it a tug to confirm it is well seated.

Charge Controller Battery Negative to Bus Bar

Connect the other end of that black wire to the charge controller battery minus terminal. Insert it fully and tighten the screw. Give a tug to confirm it is seated.

Battery Negative to Charge Controller

Take a 22" #10 red insulated wire and strip ½" of insulation off one end and ⅜" insulation off the other end. Feed the end with the ½" of insulation removed through the side strain relief. Neatly feed and bend it up so that the conductor can be inserted into the top of your 30A charge controller circuit breaker connection. Very important this wire is connected to the 30A charge controller breaker. That breaker will have the 12" red jumper wire that you attached to the bottom connector which came from the battery positive (bottom post of the 175A main inverter breaker). Insert the wire fully and with a Phillips screwdriver, snug the wire in place. Give a tug with a pair of long nosed pliers to confirm it is seated.

Charge Controller Battery Positive Wiring to Circuit Breaker

Connect the other end of that red wire to the charge controller battery positive terminal. Insert it fully and tighten the screw. Give a tug to confirm it is seated.

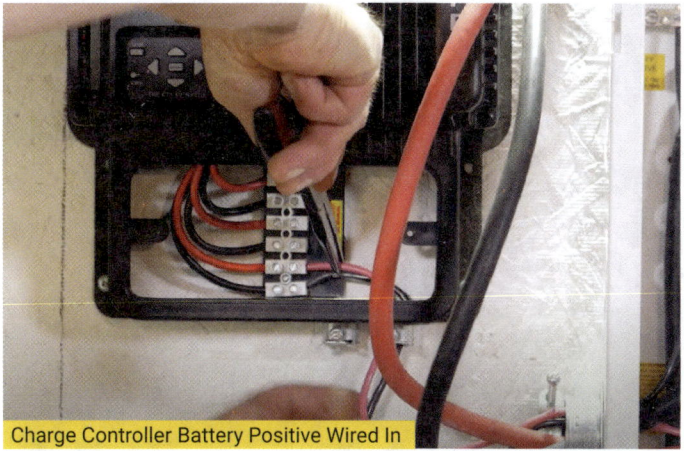
Charge Controller Battery Positive Wired In

Take a 24" #10 white (red insulated wire is OK too) insulated wire and strip ½" of insulation off one end and ⅜" insulation off the other end. Feed the end with the ½" of insulation removed through the side strain relief. Neatly feed and bend it up so that the conductor can be inserted into the top of your 20A solar breaker. Insert the wire fully and with a Phillips screwdriver, snug the wire in place. Give a tug with a pair of long nosed pliers to confirm it is seated.

Solar Panel Positive Wired to Circuit Breaker

Connect the other end of that white or red insulated wire to the charge controller solar panel plus terminal. Insert it fully and tighten the screw. Give a tug to confirm it is seated.

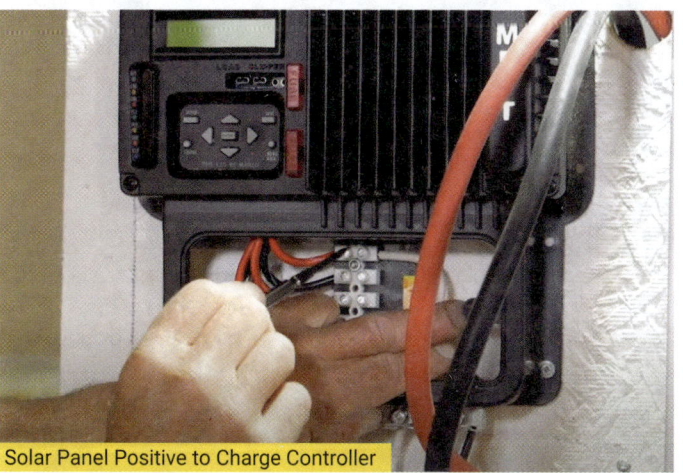
Solar Panel Positive to Charge Controller

THE MODULAR BACKYARD POWER PLANT

Just to fully explain, I like to have all black insulated wires as denoting ground or battery negative. Red insulated wires would be battery positive. My preference is to have a white insulated wire for the solar panel positive just to make it easier to see visually that the solar panel positive connection is going to the right places. It's just easier to double check and troubleshoot your wiring. But the electrons flowing through the circuit don't care what color the insulation is so if red is all you have, that will be fine.

Go back through and double check that all connections are secured. Use a wrench when appropriate or screwdriver to snug everything up. That wasn't too bad, was it?

This is as good a time as any to install a couple of cheap handles to make this easier to lift and move around. Plus, handles make it even easier if you find the bulkiness a bit of an issue.

I picked up a couple of cheap handles at the hardware store. They were installed on the back side with the inverter. Top left and top right is where they were installed.

Installing a Couple of Handles

Wiring the Power/USB Panel

You have a number of choices with this. You may decide you already have some kind of emergency lighting and have no need to wire in a special DC rated LED light bulb. Or perhaps you have no devices that would need a USB charger. Or you might opt to plug your USB phone or laptop charger into the inverter outlet and will be content to charge your device whenever the inverter comes out of power conserve mode.

But in case you need a means of light and/or a way to recharge devices, I covered all bases with this one panel. This panel provides two USB ports along with a cigarette lighter socket that you can wire a light into by adapting and wiring a cigarette lighter plug.

Power/USB Module

If one DC power outlet is not enough and/or the two USB outlets are always recharging devices, you can always add another panel to recharge more devices or run another light. You may even find a panel for sale with two power ports along with a couple USB ports. That would work just fine. Make sure any panel has a fuse inline for safety.

There are also individual USB components and cigarette power sockets you can buy. You can install more of each of them than I have installed. You have a great deal of flexibility and it is purely based on your specific needs. You are also free to bypass this panel completely as long as you understand the inverter takes 100 watts to come out of standby and a phone charger or LED light will not be enough to wake the inverter.

There is nothing special with where I chose to mount this panel. You are free to mount the panel as you see fit or you can mount it as I have done. You want to make sure the panel is easily accessible and when you apply pressure to plug a device in, the panel doesn't break off and flop to the floor, which is why I chose to flush mount it on our main board.

Here's how I mounted our panel. I found a clear spot on the front of our panel 3 ¼" from the side and 16 inches from the top down. I used a square and about 4 inches from the edge, I made a mark. I measured from that mark 2 ⅛" and made another mark. That's 2 ⅛" center to center and will be where I locate my hole saw. I used a 1 ⅜ hole saw and bored 2 holes in line with each other but spaced apart so that I had 3 ½" edge to edge. Remember, when using a hole saw, cutting from both sides gives the cleanest cut.

Then I used a straight edge and magic marker to connect each hole.

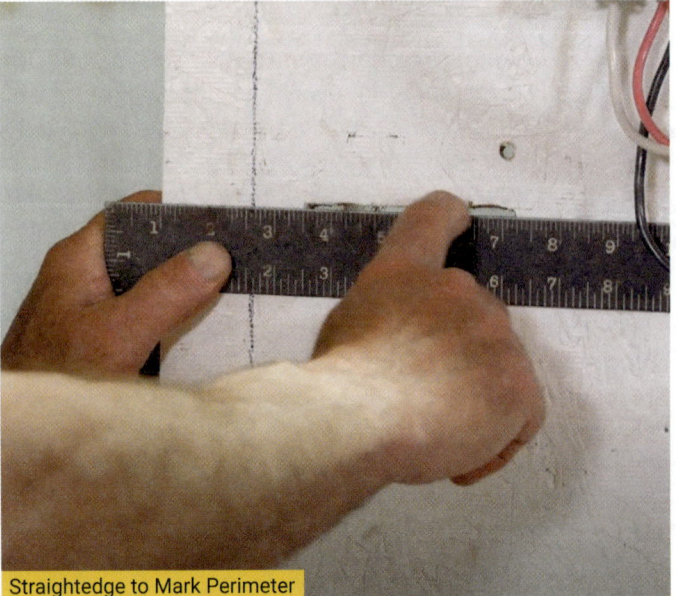
Straightedge to Mark Perimeter

I used a jig saw to cut on my marked lines and removed the piece of wood between the holes. A hand keyhole saw would work fine as well. You now should have a slot with rounded ends. You should be able to set the panel in the hole but it may not sit flush. I used a rasp file to tweak my slot so that my panel sits flush on the surface. Then I used the 2 supplied screws to fasten the panel.

Power/USB Module Mounted Flush

There are only 2 wires which need to be connected. Can you guess at this point what they are? Right, positive and negative. The negative needs to go to system ground, which is battery negative. The positive will go to battery positive. The panel I am using has an internal 16A breaker with the on/off switch so there is no need for an inline fuse. It is also pre-wired and ready to connect. However, if you buy a different panel or you get a USB and cigarette socket in individual pieces, you should install a 15A inline fuse if there is no breaker or fusing with your panel. That way it will be electrically protected. All the ones I've seen online use a spade lug. You would need to buy the female spade to crimp wires for your connections. Very easy to do.

We'll use a piece of black and red insulated 16-gauge wire cut to 32" to make our connections.

Wiring the Power/USB Module

Twist those two wires together so that they act as one wire. It will be easier and neater to run the wires this way. Feed that twisted pair through the strain relief of our breaker box so that we can make our DC connections.

THE MODULAR BACKYARD POWER PLANT

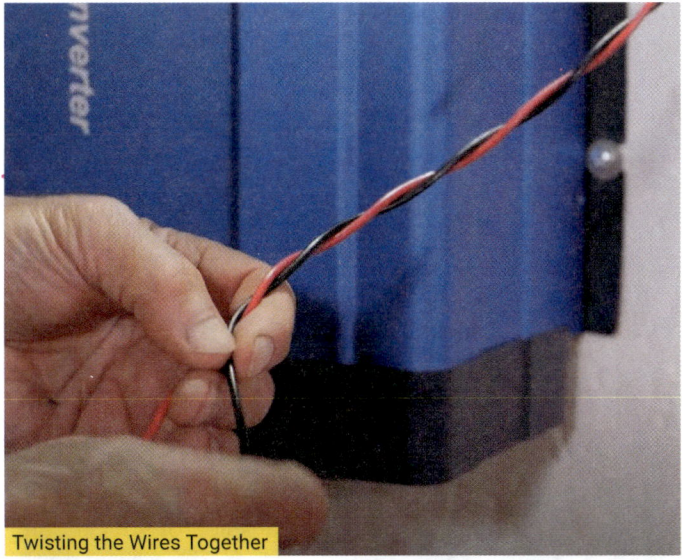

Twisting the Wires Together

Using Wire Strippers

Strip ½" insulation off of the red wire. You will need to crimp another lug terminal on to this wire.

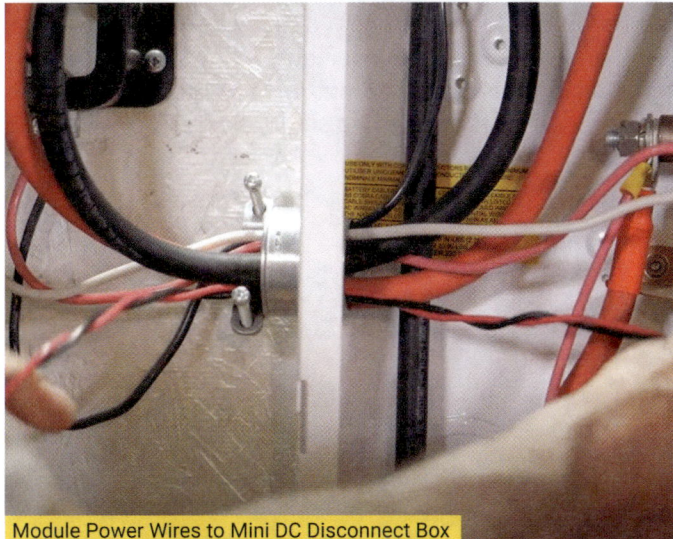

Module Power Wires to Mini DC Disconnect Box

Inside your breaker junction box is a ground bus bar where multiple DC negative connections can be made.

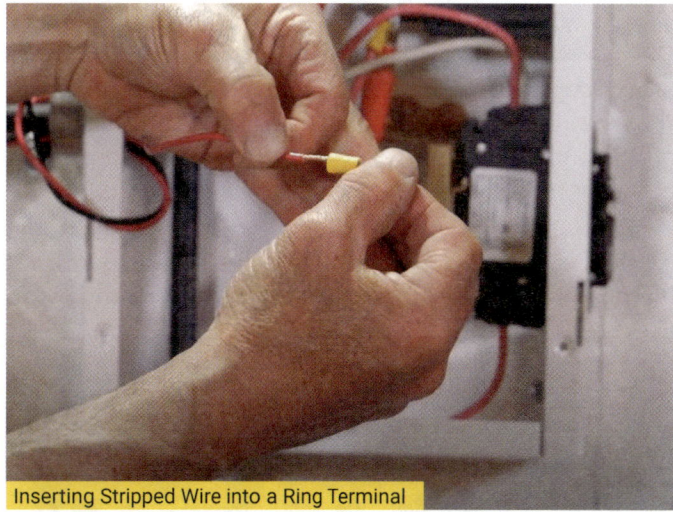

Inserting Stripped Wire into a Ring Terminal

Make sure you give the crimper a good snug.

Module Negative Wired to Negative Bus Bar

Strip ½" of insulation off the black wire so it can be inserted and screwed into one of the many ground slots on the common bus bar.

Crimping the Ring Terminal

Find the large inverter circuit breaker. The bottom connecting post has the battery positive cable and a smaller red wire you previously connected. We will add this ring

49

THE MODULAR BACKYARD POWER PLANT

lug onto that post by unscrewing the nut and removing the washers. Slide the ring lug on and put the washers and nut back on in proper sequence.

Module Positive Wired In

Tighten the nut. You have the connections made from the battery and now all you need to do is connect the other end of the twisted pair to your panel.

By the way, as long as your batteries are connected, it doesn't matter whether your whole system is off, your USB panel will be powered so keep that in mind. That may prove helpful in saving energy. In other words, you could shut the inverter off at the big breaker and shut the power off to your charge controller yet you can still charge your phone or run a light direct from this panel.

I'll show you two ways to make the connection from the twisted wire pair to your USB/Power panel. Neither method requires any special tools. The first method is to use a 2-wire push-in connector.

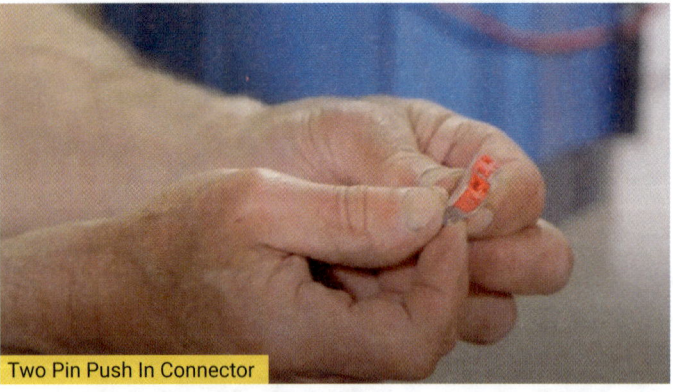
Two Pin Push In Connector

You would strip insulation off the red wire of your twisted pair and push it into any one of the openings of your connector and then push the red wire from your panel into the remaining free opening. Repeat this for the black wire.

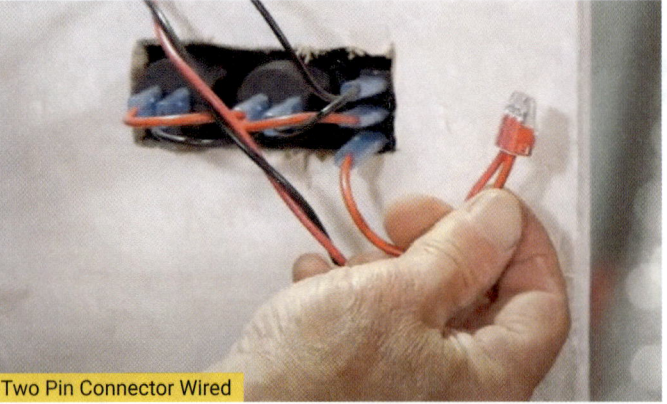
Two Pin Connector Wired

Just that easy, you've made the connection. Give each wire a gentle pull to ensure they are firm. Not too hard though or you risk pulling your wire out.

The second method is to use a 2-position terminal strip.

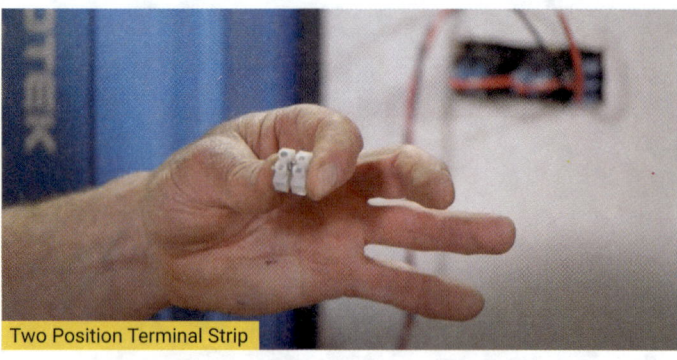
Two Position Terminal Strip

Mount the terminal strip on the backside of your mounting board nearby to the panel's two short wires. Strip the red and black insulation off your twisted wire pair. The black and red wires are inserted on one side of the terminal strip. The panel red and black wires are inserted and secured on the opposing side making sure the colors match.

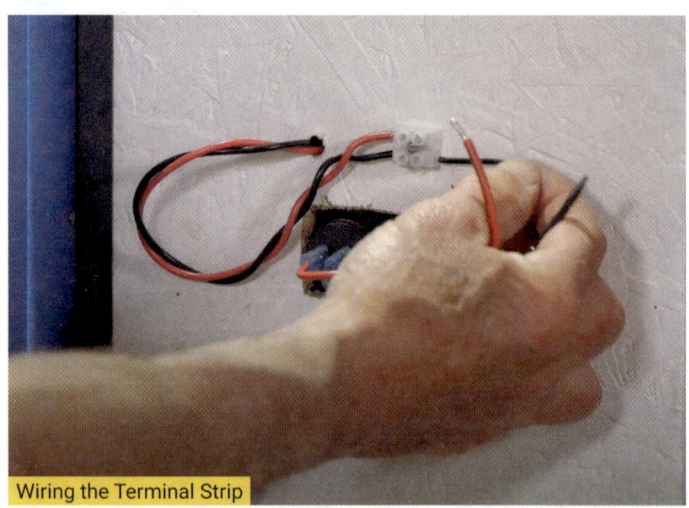
Wiring the Terminal Strip

Give all 4 wires a pull to verify they are clamped securely.

Wiring a Light Fixture with Bulb

Since you may be using a DC light bulb direct from the battery as your primary light source, let's figure out how you will wire this into your system. Bear in mind, you can certainly plug your standard AC light into the inverter which will work fine as long as the inverter is not in power save mode. If the inverter is in power save mode which is what we have set it up for, a light alone will not be enough of a load to bring the inverter out of power save mode. The light will not receive any power until more load such as a refrigerator coming on wakes the inverter and then when the refrigerator turns off, the light will go off as well.

I bought a cheap utility light with a clamp to use with our DC bulb. We will need to cut the plug off and rewire the light fixture to make it compatible with a cigarette lighter socket so make sure you are comfortable modifying an existing light fixture. Perhaps you can find a cheap light fixture at a flea market or on sale. This light fixture will be dedicated and used ONLY when your grid power is down and you need some reliable light in the house. So please don't start modifying your good expensive fixtures.

Assuming you've found a cheap light fixture that can be set up in the event you are using your modular backyard power plant for light, cut the plug off or if you have a plug that has screw terminals, get into the plug and unscrew both wires. We want access to the wires.

Carefully cut down between the two wires to separate the individual wires. Once started, you can use your fingers to pull them apart further. Only separate the wires about 3 inches down.

Separating Cord Wires

You will now have two individual wires staring back at you. Perfect! Strip 3/8" of insulation off only one wire. Pull your DC light bulb out for a second. It looks just like any other light bulb at the base. The base has a metal shell and at the tip, there's another contact.

Typical Light Bulb Base Contacts

Those two contact points must we wired with the correct polarity. If polarity is reversed, the light will not work! The metal shell and the tip are what we need to provide electricity to. The metal shell would be connected to battery negative and the tip contact would be battery positive. Now take a look at your light fixture where the bulb screws in. Note the mating metal screw in base and at the center bottom, there's a contact as well.

Typical Light Bulb Fixture Contacts

When the bulb is completely screwed into the fixture, the connections are made and the bulb will light up if power is applied.

Now take a look at the cigarette lighter plug that we will use to power up our light. Note that there's a side contact which is battery negative and the tip contact is battery positive.

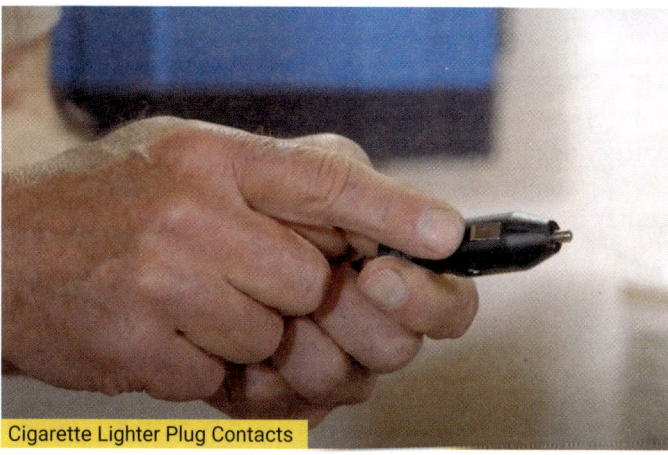
Cigarette Lighter Plug Contacts

Notice that the cigarette plug, the cigarette lighter socket, the light fixture base and the light bulb itself have the tip as the positive contact and the metal side contact is battery negative.

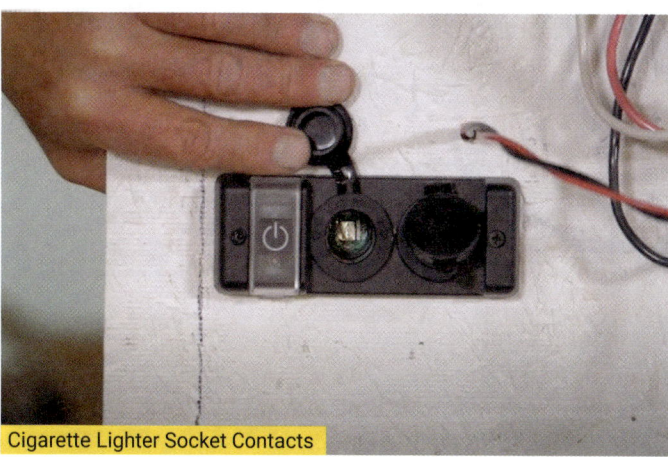
Cigarette Lighter Socket Contacts

All we have to do is ensure the negatives all contact one another and the positives contact one another and we have a lit light bulb.

But Ron, I have a light fixture with two wires and I don't know which one is the negative and which one is positive.

And I'd say back to you, that is a bit of a quandary. Seems like it's a 50-50 gamble if we just hook the wires up and hope they have the right polarity. There must be a way to figure this out and there is. Let's pull your multimeter out. This is the moment you've been waiting for. Put your multimeter on beeper or resistance.

Setting the Multimeter to Resistance

You have only one bare wire exposed. Put any lead of your multimeter on that exposed wire and hold it with your fingers of one hand. Carefully probe the light fixture contacts with your meter's other probe by first touching the metal shell. Did the meter beep or go to zero?

If it did, you have found the lead you want to connect to battery negative. Take the cigarette lighter plug and strip ⅜" of insulation off the black wire. Connect the two together by soldering or by using a connector. Remember, if this is stranded wire we are working with, we want to twist the exposed strands into a nice tight conductor. For this application my first choice would be to use a soldering iron. With a hot soldering iron, heat up and apply a little solder to each exposed wire to form a solid wire. This is called tinning the wire.

Tinning My Stranded Wire

It essentially makes the two wires much easier to solder together. After tinning, form loops on each conductor, intertwine them and with long nosed pliers, crimp the connection and then solder it.

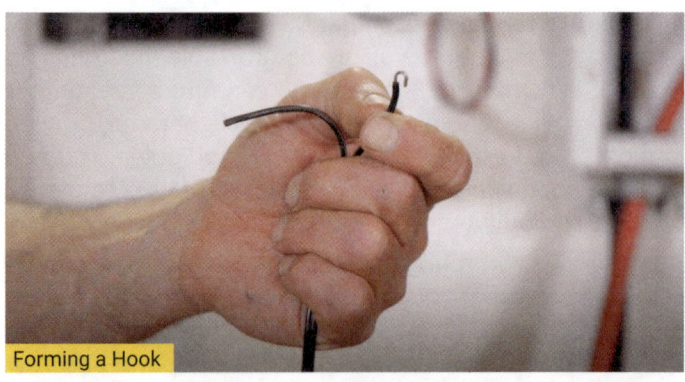
Forming a Hook

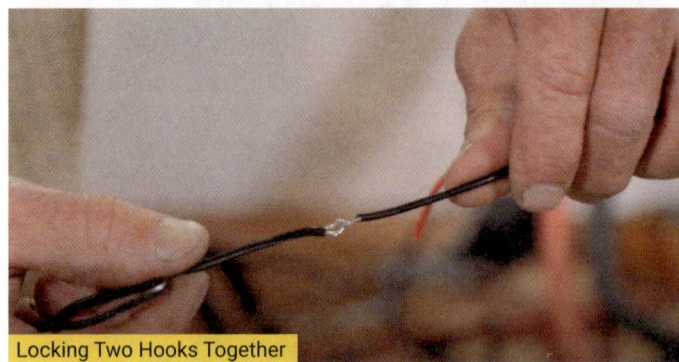
Locking Two Hooks Together

You have made a strong mechanical as well as electrical connection which would be perfect for this situation. The cord to your light will be constantly subject to movement each time the light is plugged in or moved so this requires a good strong connection. Then I would use electrical tape and make sure everything is insulated well.

Electrical Taping My Connection

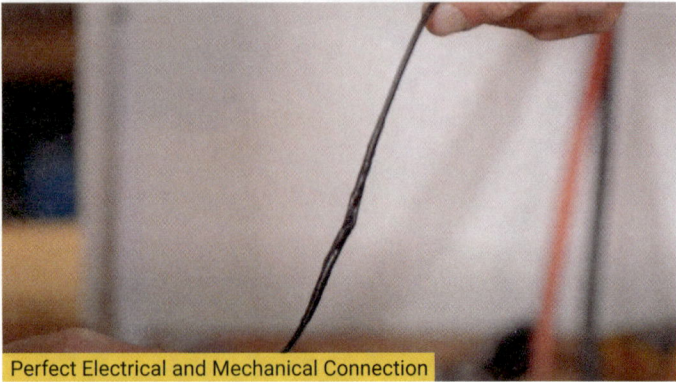
Perfect Electrical and Mechanical Connection

My second choice to connect the wires would be to strip roughly ⅝" of insulation off each wire and combine them together by twisting the pair together with your fingers. Once they are twisted together, use a wire nut to securely cover the connection.

THE MODULAR BACKYARD POWER PLANT

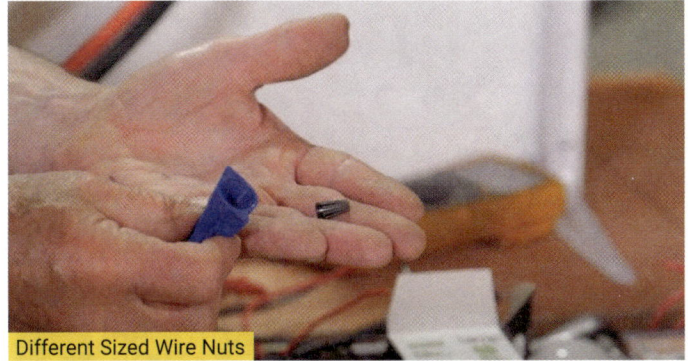
Different Sized Wire Nuts

Then use electrical tape to secure and further insulate your connection.

Electrical Tape Over Wire Nut

If you did not get a beep or zero reading on the meter, touch the center contact in the light fixture. Now the meter beeps or reads zero. You've just located the positive wire. Strip 3/8" of insulation off the red wire of the cigarette plug and connect the two wires using either of the above suggested methods.

You've now ascertained polarity and wired up one side of the light circuit. The remaining wire, whether it is red or black, can now be connected again using either soldering or twisting and covered with a wire nut and then insulated with a good tight wrapping of electrical tape.

Let's do a final check by using our meter set again to resistance or beeper. Hold one meter probe on the cigarette lighter plug tip and put the other test probe to the center contact in your light fixture. Do you have contact? Repeat by holding a probe on the cigarette lighter plug side contact and the other test probe touching the fixture socket's metal side. You should get beeps or a zero reading.

You've actually done two things with your slick testing. You've confirmed you have positive and negative battery properly connected right up to the light fixture and you've confirmed that you made two good connections. Well done!

Wiring the Solar Panels

The Modular Backyard Power Plant has been designed to work with 1 panel, 2 panels or 4 panels. It is almost impossible to get this wired wrong. The male and female plugs that come with our panels can only be plugged in one way. These are the standard in the solar industry and are called MC4 solar connectors. You should have a coil of solar wire 70' long with a male on one end and a female on the opposite end. If you ordered a longer length, these instructions still apply.

We want to open the coil and extend it out to full length. The idea here is to create two equal length pieces from this one long one. One piece we create will have a male connector and the other equal length piece will have a female connector. For our 70' piece, lets double it back on itself evenly and cut it in the middle so that you have two 35' pieces.

Evenly Divided Solar Cable Male and Female Ends

Take a look at the back of your solar panel. There is a black junction box where two wires come out. If you look on the plastic case, you will see a + and – where the wires come out of that junction box. Look at the connectors at the ends of those wires. There is a male and female connector. On the connectors themselves, there is a + and - and the positive male lead also has a red rubber ring on it which also denotes positive. The female connector would be the negative wire.

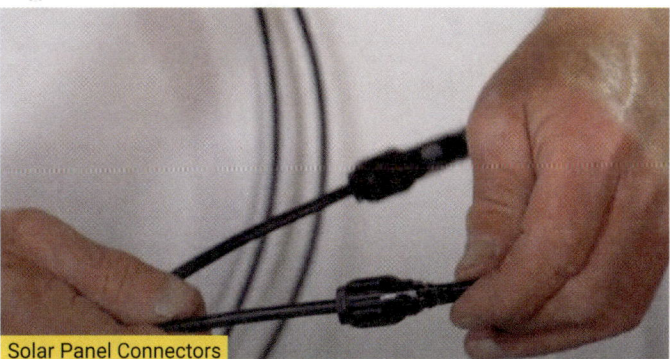
Solar Panel Connectors

With a single panel, connections couldn't be easier. Take the one 35' wire with the female end and plug into the positive male coming from the solar panel. With a black magic marker, take a piece of masking tape and label the end of that wire as "POSITIVE." Plug the male end of the cable into the solar panel's female connector and mark that wire end as "MINUS."

53

Use the decoupling tool to take those connectors apart.

Solar Cable Decoupling Tool

Neatly coil those two 35' solar cables up and you will wire them into your charge controller. Find the two ends you labeled positive and minus with tape and marker. Carefully strip ½" of insulation off each cable end with a sharp knife. You do not need to cut all the way down to the conductor itself. Just score the circumference of the cable.

Scoring Solar Cable

Wiggle the insulation and pull it free. If you need to score slightly deeper that's fine and then try wiggling again. The insulation will separate allowing you to pull it off. Remember you don't want to score the wire. You should have two cables with an exposed ½" conductor. Route them through the bottom strain relief into our breaker box.

Routing Solar Cables into Box

The end marked minus will go back out through the side strain relief directly up to your charge controller PV minus terminal. Slide the wire into the PV minus terminal connector and tighten the screw snug. Give it a tug to make sure it's secure.

Solar Panel Minus Connected to Charger

The end marked positive will go to the bottom connection on your 20A solar breaker. You already have a white or perhaps red wire coming off the top of that breaker going to the charge controller solar plus input. Again, slip the bare end into the connector and tighten the screw snug. Give it a tug to make sure it's secure.

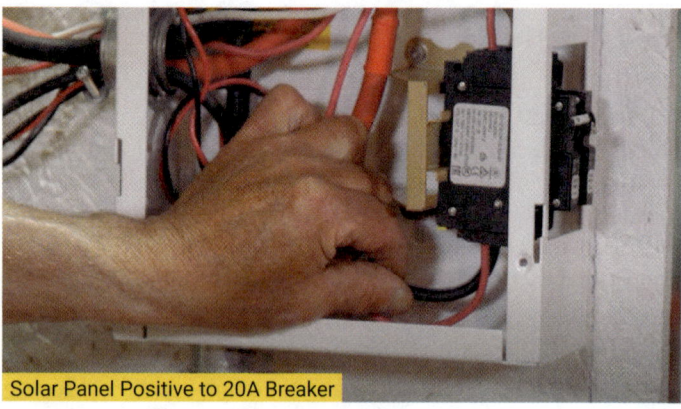
Solar Panel Positive to 20A Breaker

Keep the cables coiled and neat. The wiring for your single solar panel is complete. The ends of the solar cable with the plastic connectors will feed through an open window to connect with your solar panels and provide the solar power to your charge controller. Your single panel is now ready to be used.

If you have two panels, you will wire them in series. You can refer back to the "Connecting Batteries" section in a previous chapter to the illustration titled Battery or Solar Panel Connections. It might be easiest and less confusing if you could take them outside or carefully lay them face down on the floor protecting the glass front panel with cardboard or throw rugs. You want to orient the two panels lengthwise but with the plastic junction boxes close together.

This is the way you would set them up in a real emergency. Plug one panel into the next panel. You only have two ways to make that connection and it does not matter which way you choose. It will always be right. You can plug the positive of the first panel into the negative of the second panel or you can plug the negative of the first panel into the positive of the second panel. Either way, you will be left with a single positive connector and a single negative connector. The coiled cable with your power plant will

THE MODULAR BACKYARD POWER PLANT

Proper Panel Orientation

Care must be taken that the set of panels in the front does not shade the rear string. As well, make sure each set of panels is oriented exactly the same. For example, when viewing all the panels from the front, the two male connectors are on the right-hand side and the females are on the left, or vice versa. The females could be on the right, males on the left.

What you want to do now is parallel those two strings so that once done, you still end up with only two connections to be made to your charge controller.

We will do this using two "Y" connectors. They can only be hooked up one way. Find the "Y" connector with 2 female ends/ 1 male end. Go to the solar panels with the side having 2 male plugs. Make the connections to the "Y" connector.

Repeat on the opposite side with the other "Y" connector. That "Y" MC4 connector will have 2 male ends/ 1 female end. Make those connections.

still feed through the window and connect to your series wired panel the exact same way it would if you only had a single panel. Slick huh?

So how does one hook up 4 panels? This is actually very easy. Refer to the image for visual clues. The first thing you will need to do is repeat the above procedure for each set of two panels. In other words, you will have a set of 2 series wired panels with a plus and minus connector available and a second set of panels in the exact same configuration also with a plus and a minus connector available. In technical jargon, you have two strings of solar panels. Ideally, when you set these up on the lawn, they will be arranged with a string of panels at the back and another set of panels directly in front.

SERIES / PARALLEL PANELS WITH "Y" CONNECTORS

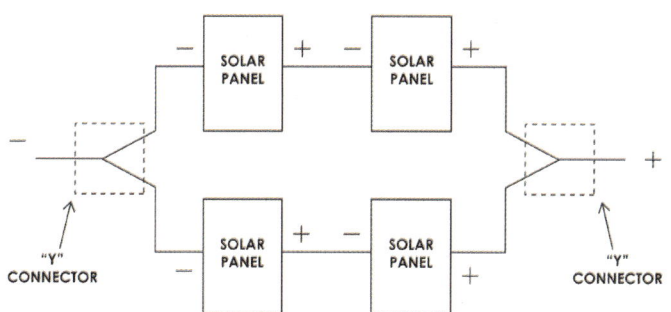

You have now utilized two "Y" connectors to make the parallel arrangement with your solar panels. As you can see, it's hard to get this wrong. You now have a male connector available on the one side of the solar array and a female on the opposite side. When it comes time to hook this up to use, you will take the cables coming through the window from your power plant and make the connections to those two "Y" connectors. Again, there's only one way they can be connected.

That's it!

You have utilized four solar panels to create a series/paralleled array. Our power plant should now be wired and ready to use.

Four Solar Panel Array

CHAPTER 8
System Start Up

Outstanding Job! You sure have come a long way with this. Regardless of how tech savvy you were when you first opened this book, if you've built a power plant following all the directions, you are amazing and it's time to make this puppy run.

The very first thing you need to do is envision that the power just went out. Where in your home would you like to place the power system and battery? Where's the refrigerator or large device that an extension cord may need to be routed to? Is there a window nearby that you can feed the panel cables through to the south facing solar panels? Take your solar panel/s and set them up outdoors in your chosen location. Move the power plant and battery to the best spot that makes the most sense in your house.

Before you connect your battery and solar panels to the power plant, let's do one more run through of all the wiring. No matter how many times in my life I've wired up stuff, I take a step back and with all my concentration, I go back and confirm every connection. This is where we want to find an oops, a wire that is in the wrong place. This is the procedure I used myself when I built my own Modular Backyard Power Plant.

I visualize in my head and talk it out loud how the current flows through the system. I also physically reach in and give each connection a tug as I go along to confirm I've tightened and secured each wire. OK, the negative lead of the battery must be connected to chassis ground which ultimately goes to inverter negative. The positive of the battery must go through the big breaker before it goes anywhere else. Then power comes off the big breaker and goes to the inverter positive. So far so good. Looks like I have that right.

I took another wire from battery positive and ran it through a smaller circuit breaker which goes to my charge controller battery positive. That breaker is well marked "Charge Controller" so if I ever need to power down the charge controller, I merely flick the breaker off. I need to make an important point here regarding the charge controller. There is a sequence that we must follow when we power up and down. We do not want to power down the charge controller with the solar panels still energized and feeding the solar charger. In other words, it is possible to do some damage to controllers if the solar panels are on providing solar energy to the controller but the controller itself is turned off at the breaker and itself is unpowered. Some charge controllers including this one may have protection, but it's just a good habit to get into. I'll mention this a couple of times in this section.

Battery negative also needs to go to my charge controller's battery minus terminal. Ok, that looks right. The battery is properly connected to the charge controller so the charge controller has power.

I was careful and double checked that my solar panel/s are all connected properly and are set up outside facing the sun. Since this is only an initial start-up and test, I am only concerned at this point that the solar panels are in sunlight producing some energy. If this was a real grid down situation, you'd want to make sure your solar panels are located so that at no time during the day will shadows of any kind fully or partially block any of your deployed so-

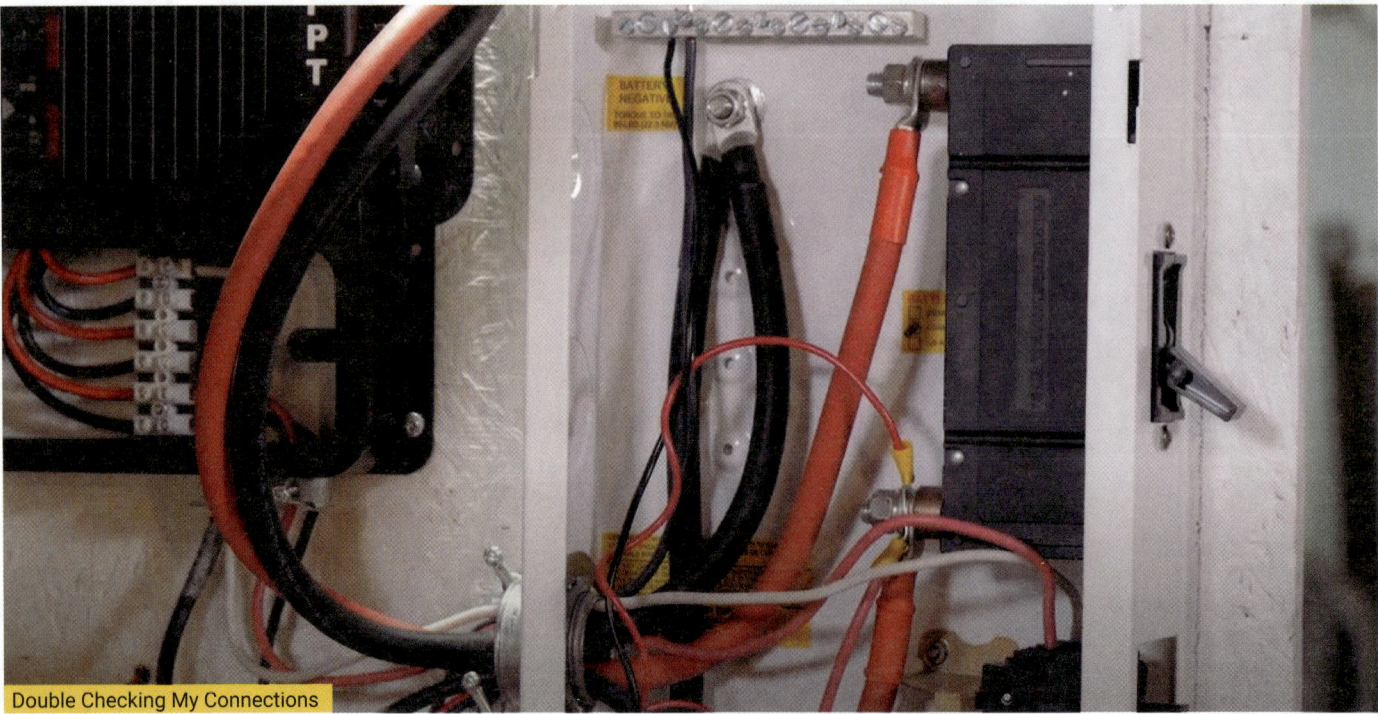

Double Checking My Connections

lar panels. It is imperative every panel see full sunlight throughout the day. This will be the time to assess the placement of your panels in relation to the tracking of the sun. Bear in mind, worst case scenario will be around Dec 21 when the sun will be the lowest in the sky. But regardless of season, this will be a good gauge of how good your panel location is.

Next, I check to be sure the negative of my solar panels goes direct to the negative PV input of my charge controller. Because this Midnite "KID" solar charger specifically states to keep PV negative input and battery negative completely separated, I've confirmed I have not tied those two points together.

I want to be able to protect and safely turn off my solar panels if I want to so my positive lead of the solar panels goes through another small circuit breaker which is also well marked with "Solar Panels."

Two Solar Panels Connected in Series

As well, referring back to the previous chapter and the solar panel wiring illustration, make sure that if you have multiple solar panels, they are wired together properly.

Two panels are simply wired in series and if you have the full 4 panels, each set is wired in series and then those sets are paralleled through the "Y" connectors.

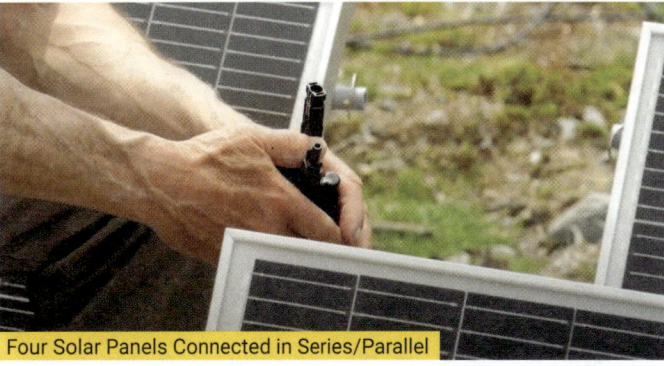
Four Solar Panels Connected in Series/Parallel

You should only have 2 free wires available to make your connections to the system. One male connector and one female connector. They can only be connected one way.

Circuit Breakers Well Marked

All Four Panels Connected

The output of the breaker goes to my PV positive terminal. I'm convinced I've wired this up right so far. I've verified that my solar panel cables are properly wired to the input of the charge controller. So far, so good!

The last piece of the puzzle is my little USB/power module. Negative of that module simply goes to ground, battery negative. That makes sense and I've supplied the positive 24 volts through an inline fuse so it's protected. You should have gone through the same procedure as I did.

If you have more than one battery connected, you just need to do a double check to be sure the batteries are in parallel. Those battery connections were made using the short 10" #1 AWG interconnects. Positives connected to positives and negatives connected to negatives.

In 9 paragraphs, I just described the total wiring of our power system. I've confirmed all my wires are right.

We are so close to making things play. But we need to curb our enthusiasm and turn everything on in a proper sequence. Before we hook up our battery, let's make sure our Main inverter breaker is OFF. The two smaller breakers, charge controller and solar panels should be OFF (The levers of all circuit breakers should be flipped down to the "OFF" position) and the inverter rocker switch is OFF in the center position.

Just to be absolutely clear here, I am talking about the three circuit breakers you wired in to the electrical box of your power plant. We are not anywhere near your home's electrical circuit breaker panel.

THE MODULAR BACKYARD POWER PLANT

Circuit Breakers OFF!

When I connect the battery, I do not want to energize any of our components. Next put your safety glasses/goggles on. The battery negative lead which we marked battery negative can now be connected to your battery negative terminal. Make the connection snug and tight but do not over tighten. If you have a torque wrench available, by all means use it and tighten the battery bolts to 88.5 to 106.2 inch/pounds as specified by the manufacturer.

Connecting Battery Negative

Find the unconnected positive lead that you need to connect to battery positive. I like to take the cable and for a fraction of a second, just touch the cable to the battery terminal. Since all breakers are off and there's no electronics getting power, nothing should happen. No small spark and certainly no large spark. This is just a double check test I do to make sure something isn't wired radically wrong. If you had a big spark, you have something wrong and the problem needs to be found before you go any further.

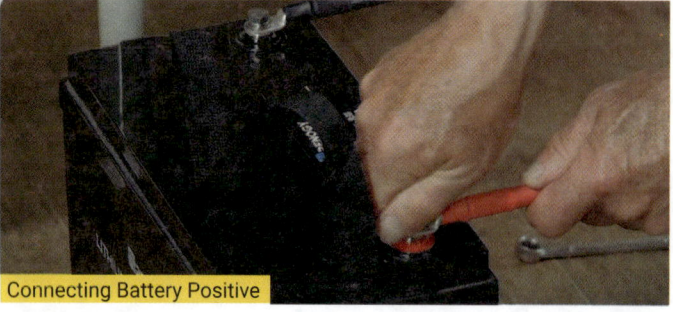

Connecting Battery Positive

Once you are satisfied all is well, make the connection to the battery positive. This connection is straightforward if you only have one battery. However, if you have more than one battery hooked up, the best connection for multiple batteries is to connect the negative as usual but go diagonal to the opposite corner of the battery bank to connect the positive cable.

Diagonal Battery Connections

Remember, these batteries can only be hooked up in parallel. Positive terminals connected together and negative terminals connected together.

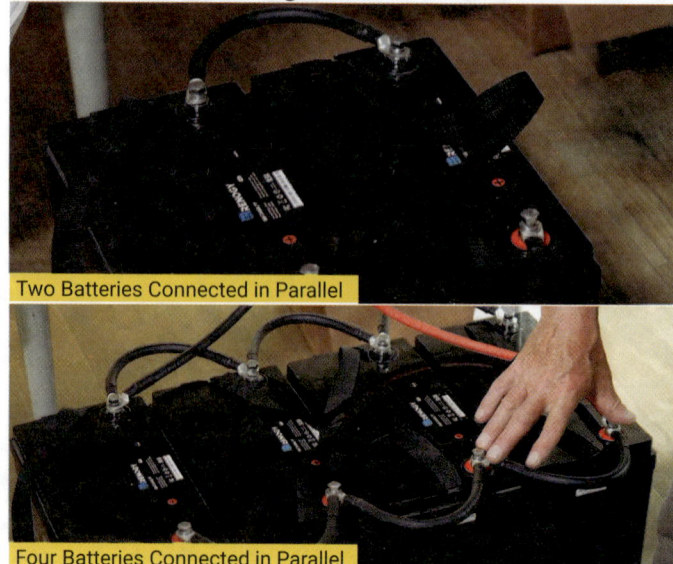

Two Batteries Connected in Parallel

Four Batteries Connected in Parallel

Again, use a torque wrench to manufacturer's specs if available. If not, snug tight but do not over tighten. This is the time to triple check that the black battery cable is connected to battery minus and the red battery cable to battery plus.

Now open a window a crack and feed your solar panel cables out the window.

Solar Panel Cables through Window

Head outside and push the mating connectors together for each pair of wires.

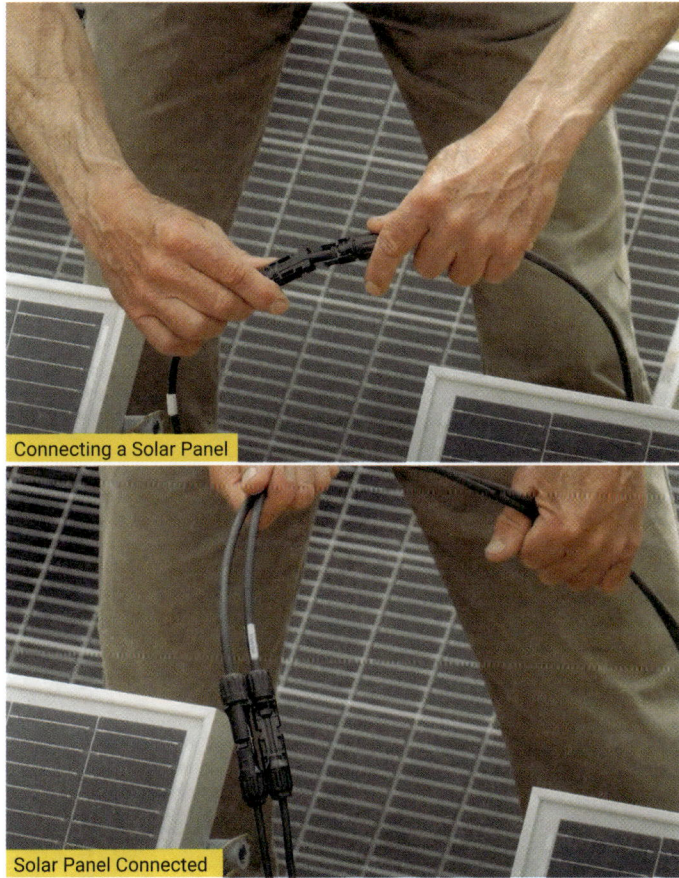
Connecting a Solar Panel

Solar Panel Connected

Remember you have the plastic disconnect tool that makes taking those connectors apart a cinch when it's time to store everything away. The system battery and solar panels are now wired in but nothing yet has been powered up.

Head back inside to your power plant. Flip the large main breaker up to the ON position. That breaker supplies power to the inverter.

THE MODULAR BACKYARD POWER PLANT

Turning Main Inverter Breaker ON

The inverter now has power but the inverter rocker switch still is in the OFF position. It's time to turn the rocker switch to the ON position. Note this rocker switch has 3 positions. And if you look closely, there are markings just above the switch with on/off and an icon for "remote." Also Located on the switch itself is a "I" which is ON and "II" which is remote. We do not have a remote so we can ignore the "remote" icon. We want to turn the unit ON. Flick the rocker to ON. The inverter should come to life.

A green LED light should come on signifying the inverter is powered and ready for operation.

Inverter Full On with Steady Green LED

An orange or red light denotes a problem. See the trouble shooting section of this book and the inverter manual if an orange or red light is showing. Green is good! Steady green means the inverter is fully on. At this point we want to see a slow green blink meaning we are in power save mode. If the green LED is blinking, you did a wonderful job setting the DIP switches we modified before mounting the inverter initially. Good job!

THE MODULAR BACKYARD POWER PLANT

Let's plug something into the inverter outlet and make sure the device runs. Whatever you plug in, it needs to be at least a 100-watt load to bring the inverter out of power save mode. How about a power drill or circular saw? Maybe a kitchen appliance?

Plugging an Appliance in

If the inverter doesn't kick on, plug a couple of devices in at the same time and turn them on together. The inverter should run them both.

Now that we know our inverter works, unplug your saw or kitchen appliances so that the inverter returns to power save mode. We will turn the solar panels and charge controller on and start recharging the batteries.

Find the small breaker labeled "Charge Controller" and flip the breaker up to the ON position.

Turning Charge Controller Breaker ON

The charge controller breaker is 30A. The charge controller should come to life. Lights will flash and the display will come on.

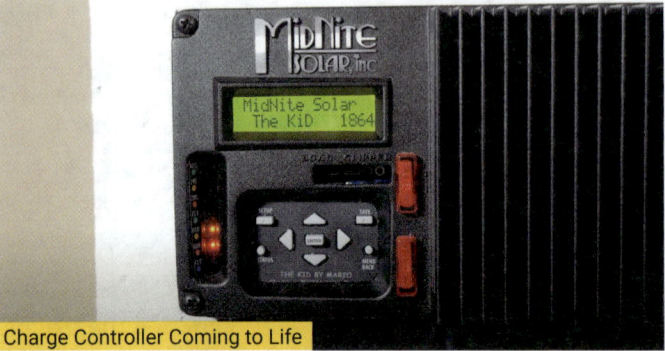
Charge Controller Coming to Life

Instead of me copying or paraphrasing the information in the manual, let's just refer back to the manual at this stage since it is well written and spells out the proper set up of this device. As well, any changes the manufacturer makes to the start-up procedure will be reflected properly if one refers back to the manual supplied with your unit. You will want to find the section titled "Initial Set-up and Use" in the manual that came with your KID. (I'll bet some of you are wishing your first newborn came with a manual) When the unit is first powered on, it will go into set up mode. This is where we can start to enter some information into the system so that it can work properly. This is where we define what our battery voltage is, what type of battery it is and what we'd like to use for charging specs.

We will set the battery voltage to 24V, the battery type will be set to lithium and the charge voltage will be 29.0 as per the manufacturer's recommendations.

Once you set up and move out of this mode, it will always remember the settings you initially made. No different than a radio. The radio has a power button with volume. You can set a station, set balance and set bass and treble. Every time you turn the power on the radio, it is preset and works the same way until you modify some setting. The same principle applies to your charge controller. Set it and forget it unless you need a reason to go back in and change something.

Press the status button to look at the display. Note the information provided. You should see 0 watts going into the system.

Initial Display with No Solar Input

Now find the breaker marked "Solar Panels" and flip that up to the ON position. That's the 20A breaker.

Turning Solar Array ON

The solar panels should be providing charging energy to the batteries. Look at the display again and confirm the system is supplying some amount of watts depending on how many panels you have set up as well as how much sun is shining on those panels. Remember each panel is 200W so hopefully you are reading a substantial wattage output for each panel you have. Ideally 200 watts but you already know that's not going to happen. Nevertheless, that's the theoretical target. Only through experience will you know what your system output will be and you can use that as a gauge any time you visually visit this display output.

Some Solar Input and Battery Charging

Well done, your solar panels are recharging the battery as it should. You've proven the inverter works, solar panels are providing energy back to the battery and you have everything wired correctly.

When it comes time to take everything apart, you will reverse the order you used when you powered everything up. Let's turn the solar panel breaker OFF disconnecting all solar input. Then shut the charge control breaker to OFF and then finally, the main breaker is turned OFF. Let's label these three circuit breakers with a magic marker in numerical sequence starting with the main breaker which would be numbered 1. The charge controller breaker would be 2. and finally, the solar breaker would be 3.

Numbering Our Breakers

That is the start-up sequence. Turn breaker 1 on first followed by breaker number 2 then breaker number 3. To shut down, reverse that order. Breaker 3 then 2 then 1. When you need to deploy your power plant in the future, you won't need to ponder how to turn it on if everything is numbered.

Congratulations!
Your Modular Backyard Power Plant is operational and you are almost ready to put it in storage. But before you put it away, let's really see what it will do for you.

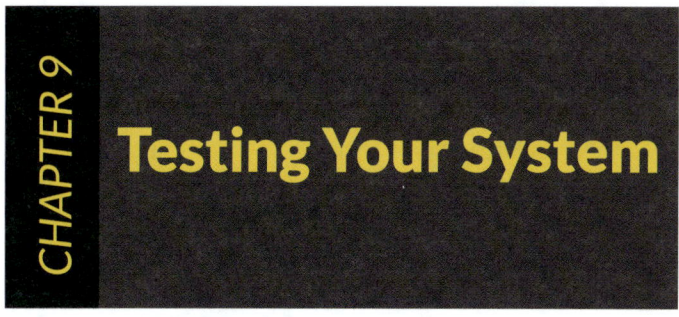

CHAPTER 9
Testing Your System

Your system has turned on and is working great. At least, we think it's working great. We know the inverter works and any gadgets we plug into it run just fine. We've confirmed the solar panels are putting energy back into the battery by viewing the information on the digital display of the charge controller.

But how does your system work given the devices you want to power? As we saw in the design chapter, there are so many variables it's impossible for me to figure out how well your system will run with your own particular devices. There's only one way to really tell.

We will need to give this a test drive so to speak. Let's set it up and see what it does in the real world with your own stuff. This will be a simulation that will give you further confidence you can run your appliances and devices for the long term should power go down. Making calculations can only go so far and this is the time to power things up for a day and confirm it does what you hope it does. Put the solar panels out in the sunshine in the morning. This is a real-life simulation so those panels need to be set up as if you really have lost power.

Go through the proper circuit breaker sequence to turn your power plant on. Confirm once again your system is running. Plug your refrigerator into one of the inverter's outlets. If the refrigerator was running when you unplugged it, it should start right back up again using your new power plant. If the refrigerator was off when you unplugged it, monitor it. It should kick on at some point and purr just like normal. Let it run for the day then assess your battery status. If you have a sunny day, the system should be running purely off the energy provided by the solar panels.

Switch on your Power/USB Panel.

Switching Power/USB Panel ON

Plug your phone or laptop in to the USB port. Preferably your phone or laptop has been discharged to a degree prior to plugging it in. No point trying to charge your device if it's fully charged to start with. Plug your modified DC light in to the cigarette power socket and turn that on at night.

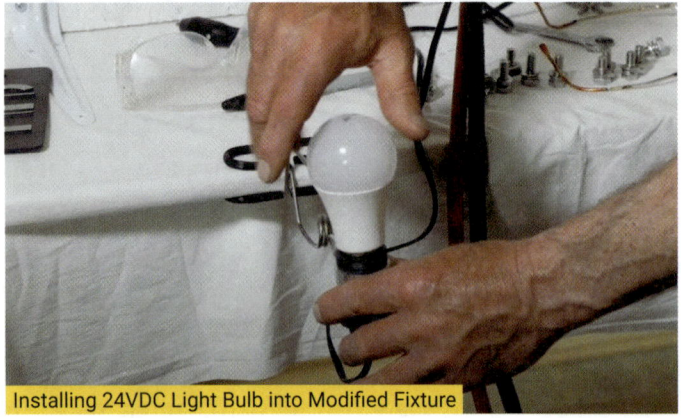
Installing 24VDC Light Bulb into Modified Fixture

Ideally, you want to keep everything plugged in overnight and let stuff run off your power plant. How are the batteries the next morning? If the sun is shining the next day, let the solar panels do their thing. At the end of the next day, assuming you had sun, how are things going? Everything ran well? The batteries fully charged or only partially? If you have matched solar panels and number of batteries to your own appliances properly, I'd expect the batteries to be fully recharged by the end of the next day assuming you received 4.5 hours of full sunlight.

You can get a sense of battery state of charge by looking at the string of LED lights on your KID charge controller. Refer to the manual for further information but there are several LED lights which will be helpful. The top green LED light will come on signifying a full charge within the last week. The 4th LED light down from the top will turn green when the KID enters float mode. Remember, the first stage is bulk where the majority of the charging is done. Then absorb and finally float which is a trickle charge to keep everything topped up. So, if you enter float, you are charged up.

Pressing the status button also provides lots of useful information. Scroll through the various screens, take a minute to digest the info and then move to the next screen. These screens and displays are adequate for a temporary power system such as this.

Do you have a device you consider higher priority than the refrigerator? Plug that in and make sure that runs. You get the idea here. The priority devices you need to keep running should all be tested. Assess how well the system powered your gadgets and whether or not you have the battery capacity and solar panels to do the job. If not, bumping up to the next module or perhaps even the Part-Time Power Plant Module should be considered.

This is a great test to figure out how you will ultimately set the system up in your house and run any wires or extension cords to their destinations. Safety is paramount here. Use good quality extension cords if they are needed. Although this is a temporary setup, you have no idea of the duration of a grid down scenario so you might as well figure out how you will power your things with extension cords for a lengthy grid down situation. For safety and efficiency reasons, any extension cords used should be the proper length to go from the power source to the appliance. In other words, if the appliance is 25 feet away, a 150-foot extension cord is a poor choice to make the connection. Get a proper 30' or close heavy duty extension cord and be safe. Run the extension cords with care so they do not cause someone to trip. Traffic areas might need to have the cords in those areas taped down. The last thing we want to see is someone catch a foot, lurch and fall head first into the toilet.

The same goes for the two wires from the solar panel. They will need to come into the house so my suggestion is to locate the panels near a window remembering the panels must face the sun and not be in any shade. That way all you have to do is open the window a crack and slide the wires out to the panels. You can stuff a towel in the crack to seal the opening. That towel will keep bugs out in summer or keep heat in during the winter.

If you are concerned about your panels disappearing or your personal safety, carry them into the garage, shed or house for the night and reset them at sun up. Use the plastic decoupling device to separate the connectors. Don't forget to bring the wires in for the night, close and lock your window. Otherwise, just leave everything setup to resume solar charging the following day.

It's time to wrap this project up by replacing all our covers. Make sure all circuit breakers are Off. Disconnect the battery. Disconnect the solar panels. Before we put the covers back on, let's once more go to every connection and make sure they are all snug and tight. There's a tendency for things to settle in place and you will probably find a screw or two can be snugged a tad, especially with stranded wires such as the solar cables.

Put the black plastic cover on your charge controller. Next, there are 2 metal plates that came with your DC mini disconnect box. Find the plate with the 5 tabs discarding the plate with the 3 large tabs as it will not be used.

I'd suggest removing tabs 2 and 4 on the plate with the 5 tabs. Use a screwdriver to pry one corner out and with pliers work it free.

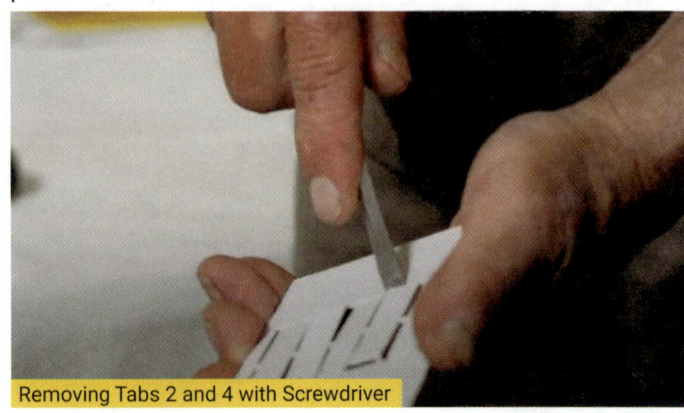
Removing Tabs 2 and 4 with Screwdriver

THE MODULAR BACKYARD POWER PLANT

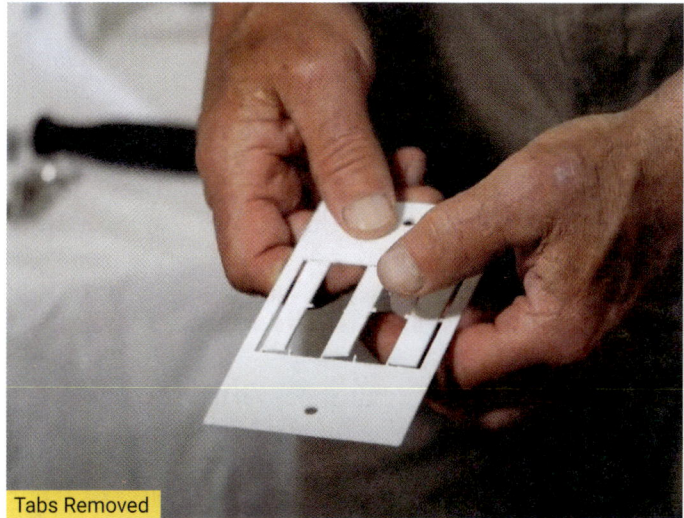

Tabs Removed

This plate is the cover to the two smaller circuit breakers. (Solar charge controller disconnect and solar panel disconnect) Use a flat file to smooth any nubbins from the removed tabs.

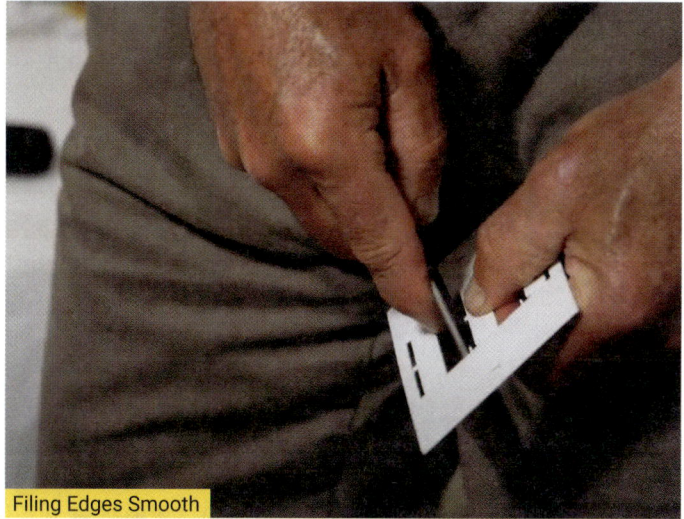

Filing Edges Smooth

With 2 screws that came with your DC mini disconnect box, position the plate over the two breakers and fasten the plate in place. If need be, gently slide the breakers to align with the plate openings.

Plate Installed

And finally, slip the remaining large cover on the DC mini disconnect box and screw it in place with the 4 remaining screws.

Installing Final Cover

I want to leave you with a couple last thoughts here. The system is no good if it isn't ready to go at a moment's notice. That means having the batteries charged. If you have any advance notice of impending weather or potential power outage, take advantage of that time to fully charge the batteries using the grid power and battery charger. Prepare your water source and emergency supplies too.

As well, protect your Modular Backyard Power Plant in the safest place you can think of. After all, you've worked hard on this and expended money just for this particular scenario. The power system can't help you if it was blown into the next County or damaged in some fashion. You may end up with a damaged home but you are safe and you have a means to power tools, phone, lights and what have you. You should be in much better shape than the majority who didn't have the foresight you had in your preparations.

Remember, your mental attitude and how you approach the disaster will have a large bearing on how well you deal with problems. Take those problems one at a time so they don't overwhelm you. Prioritize the most important things to deal with initially.

If by chance, power gets restored whether you expected it or not, don't assume it will stay on. Plug that battery charger into the grid power pronto and top up the batteries. In the event the power goes back off, because you anticipated this possibility, you are back ready to deal with whatever comes your way. And if the power stays on, well done, you were awesome in the face of adversity with far more confidence than you started with.

When it comes time to safely store your power plant, remember to power down in the proper sequence. Don't forget to put those safety glasses back on when it's time to disconnect the batteries and put everything away.

CHAPTER 10

Troubleshooting

If something doesn't appear to be working properly, before we hit the panic button, let's do some thinking together on this. If the inverter won't turn on when we flip the main breaker and the inverter rocker switch to the ON positions, let's double check things. Are you sure you are flipping the big 175A breaker to the up position? Are you sure the little rocker switch on the inverter itself is toggled to the ON side? Remember, if you toggle to the wrong side, that is the "remote" input and it won't work. The switch should be depressed to the side labeled ON (I).

Confirm your battery is properly connected. The red battery cable coming from the large 175A breaker should be going to the positive terminal of the battery. The black cable attached to the post welded to the chassis should be going to the battery negative terminal. A heavy black cable comes off the same welded chassis post and goes to the inverter negative battery terminal. A heavy red cable comes off the 175A breaker and goes to the inverter positive battery terminal.

With your multimeter, check to make sure your battery voltage is above 24VDC. Set your meter to read DC volts. The black test probe goes to battery negative and the red probe goes to battery positive. If the battery voltage is not over 24 volts, the battery either was never woken up with an initial charge or it shut itself off. Make sure the main inverter breaker is OFF. Turn the other 2 breakers OFF. To be completely safe, let's go out to the solar panels and disconnect any one lead from the panels (doesn't matter whether positive or negative lead), so there is no chance of current flowing into our system. Put your safety glasses/goggles on and disconnect the positive cable of the battery. Recharge the battery with the charger until the charger determines it's fully charged.

After the battery has charged, confirm at least 24V with your multimeter, don safety glasses and then re-hook the battery positive. Flip the main inverter breaker on and then turn the inverter toggle switch on. Does the inverter turn on? If yes, it was just a weak or dead battery. If it still won't turn on, check the battery voltage again with your multimeter. With the main inverter breaker on, go to the inverter input terminals and confirm you have the same voltage there as you just measured at the battery. You should have at least 24V at the input to the inverter. If you do not, trace back your wiring to the battery. It is only 2 wires going through a big circuit breaker. With your multimeter's black lead held on battery negative, you should be able to use the red test lead of your meter and probe different points of your battery positive's wire runs.

If you have battery voltage at the inverter and the inverter still won't turn on or it does not function as expected, again double check that you are flipping the rocker switch to the on position and not the remote position. As well, double check the DIP switches. Carefully go over them one at a time. You may have inadvertently set it to remote or other setting by mistake. If you are convinced everything is right and it still won't work, it may be a bad unit. Give tech support a call where you purchased the inverter and see what they suggest.

If the inverter came on but the inverter LED is flashing red or orange; the flashing orange LED denotes a warning. With an orange LED flashing either slow or fast, the inverter is sensing the battery voltage is either getting too high or too low. Refer to the user manual. Unless the charge controller somehow stopped functioning or was set for too high a charge voltage, I don't know how you would get the battery voltage high enough to trigger a warning or shutdown. But too low a voltage means the battery has been discharged too far and the voltage is too low to safely power the unit. Recharge the battery before going any further.

A red LED signifies a major problem has been detected and the unit has shut itself off for protection. Let the unit cool down for a spell and retry. Refer to the manual as to possible causes. Are you trying to power loads over 2500 watts at the same time? That will certainly shut the unit off.

Lastly, if the inverter won't come on even though the green light is flashing and something has been plugged in, the load may be too small to trigger the inverter to come out of power save mode. Plug something else in to provide a load over the 100-watt threshold.

Do you have a problem with the charge controller and/or solar panels? The first thing we need to be aware of is this charge controller has double fuse/breaker protection. You will see three slots on the face of the KID controller. Two of those slots have a removable red plastic cover with fuses for Solar and Battery input. In addition to those onboard fuses, Midnite recommends additional circuit breakers. That is why we have a solar input and charge controller circuit breaker. The circuit breakers not only provide primary protection but are a means to easily disconnect the device from solar and battery. The fuses would be secondary protection.

Let's isolate and look at these one at a time starting with the charge controller. Much like the voltage measurements you took at the inverter to confirm power was right there at the inverter terminals, you can do the same thing with the charge controller. Are you sure the breaker marked "Charge Controller" is flipped up to the ON position? Take your multimeter and measure the battery voltage. Confirm at least 24 volts. Next, remove the cover to the charge controller wiring so you can access those terminals. Measure voltage between the two terminals marked battery minus and battery plus. Your meter's black lead should go on the minus terminal and the red test probe on battery positive. Again, this is a measurement right on the charge controller's battery inputs. You want to confirm battery voltage is present.

Is voltage present but the charge controller just won't turn on? Check the fuse on the front of the controller under the red plastic cap marked "Battery." Pull the fuse out to check for continuity with your multimeter. If the fuse is good, reseat the fuse and see if the unit awakens. If not, consult with the dealer or manufacturer since it would appear the unit might be faulty.

Does the charge controller power up but no solar power appears to be going to the batteries? Make sure the charge controller is powered up. Then confirm the breaker marked "Solar Panels" is flipped to the ON position. As well, this measurement must be done when the sun is shining on your panels. Measure the terminals marked PV plus/minus. You know the routine by now. Multimeter's black probe to PV minus and red probe to PV plus. Make sure you set the meter to read a higher DC voltage. You can generally expect something over 35 volts. It really depends on how much light is shining on your panels and how many panels you have connected. What is your voltage? If there is zero voltage, assuming your solar panels are in sunlight, there is something wrong with the panel wiring or perhaps a bad panel. Double check all your wiring and connections. You should be able to unplug all the panels and measure voltage out of each one while it is positioned in sunlight.

If each panel puts voltage out as measured by your multimeter, plug them back in and try again. You've proven the panels themselves are good if each has the same voltage when tested individually so there might be something funky with the wiring after they are connected together and run into the house. Maybe a bad connector? Don't forget to check the "Solar" fuse located on the front of the controller under the red plastic cap. Pull the fuse out to check for continuity with your multimeter. Reseat the fuse and see if you have energy from the solar panels now.

Does the USB/power panel work? You should be able to plug a device into the USB for charging and the light or other low wattage 24-volt gadget into the cigarette lighter socket. Your USB/power panel may have a blue LED lit signifying power is at the panel. If neither works, confirm the wiring is hooked up properly. If your unit has a switch, is it on? Has the fuse blown? Check that. Toggle the on/off switch a couple times since our particular module has a built-in circuit breaker. Use your multimeter if in doubt. Take the fuse out. Put your meter on continuity or resistance and confirm there is a connection through the fuse. Your meter should read zero or beep. If it doesn't, replace the fuse. Why did it blow in the first place? Did you try to plug too large a device in? Is it a 15A fuse and not 1.5A?

If the fuse or circuit breaker is OK, try to back off the spade lug terminals a little bit if your unit has those connections so that you can squeeze your multimeter probe tips in and measure the voltage somewhere on the outlet itself. You might need to manually wiggle the connectors a little on the back to access enough of a contact point. If there is voltage present, your cigarette lighter plug might not be making a good connection when plugged in or wired properly.

With your multimeter and a basic understanding of what voltages you expect to see, you have a good chance to cipher out any problems by logically and methodically following the wiring, taking a reading and moving on down the line until you find the problem.

CHAPTER 11
Helpful Tidbits

Believe it or not, there was a time not that long ago when the modern conveniences such as a dishwasher or clothes washer didn't exist. And yet, somehow clothes were washed and dishes were cleaned. I'm sure everybody has hand washed dishes before and one can certainly resort to that while the grid power is off. But what about some of the other things like clothes washing, cooking, water and lights?

Let's tackle clothes washing first. There are alternatives to the modern clothes washer. But there's no need to grab the clothes basket and run to the nearby stream to bang everything with rocks.

Back when I first started homesteading 43 years ago, I had a dual wash tub made for the purpose of washing clothes.

Wash Tubs and Outdoor Laundry Drying

One tub was for washing while the other was for rinsing. I'd heat hot water in a pot on the wood stove, pour it into the tubs, use a standard dedicated toilet plunger as an agitator to work my clothes in the tub and use the scrub board for stubborn dirt removal. Then when I thought they were purified enough, I'd hand wring them and then plop them into the rinse water. More plunger action and then everything was hand wrung again and then hung to dry on a clothes line.

If you don't have a couple 100-foot coils of clothesline or other reasonable strength rope, it would be a good investment. You never know when rope will become invaluable such as use for a clothesline.

I know a few of you are going ICK! about using the toilet plunger to wash clothes so I want to ease your mind. This was not a dual duty toilet plunger. It was brand spanking new and used purely for the purpose of washing clothes. But in desperation, reclaiming a contaminated plunger with hot water, soap and then sanitizing bleach will work.

A couple of plastic buckets can do the same thing as the dual wash tubs albeit on a smaller scale. Or you could use your existing utility sink with the stopper in to hold hot water. Do the wash cycle with some laundry soap and a plunger of some sort and then pull the plug and let the water drain; then just add clean water to rinse the soap out. In a pinch, there's nothing wrong with hand agitating. Just use your hands to swish and swirl the clothes around, get in a good physio workout and you have clean clothes for a while. Hanging in the sunshine outdoors will do wonders for airing out and freshening your clothes.

What is the plan for water? Do you have a reliable water source regardless of grid power? If so, you are in good shape. For those who do not have access to reliable water, if you have advanced warning of a big storm, fill up and inventory some as soon as possible. Use the bathtub and water jugs for storage. This wouldn't be suitable storage for drinking water but it would be great water for washing hands or bathing. Obviously you would dipper out water as needed and not use your whole supply for a single task.

When we shocked our well (sanitized) and knew our well was going to be out of commission for a while, Johanna filled up all the large kettles she uses for canning with water. That water was used for cooking, dish washing and so forth. She would simply ladle out was was needed for the task.

Prioritize the drinking water above all else. Be sure containers for drinking water are pure and kept separate from any contaminated sourced water.

Do you have a drilled or dug well with an accessible cap? There are devices and water well buckets that can be lowered in to a well to access water. Both commercial devices and homemade gizmos you can make yourself. This applies to drilled wells with narrow casings too. You should have some idea of where your typical water table lies. That at least gives you an idea of how far down you need to go with your bucket and retrieval line. So, for many folks, these devices might be a practical method to secure water.

If you don't have a well, you might consider rain water harvesting. A tremendous amount of water falls on a roof when it rains. Figure 623 gallons of water can be harvested from each 1000 square feet of surface area with a 1" rainfall. That's a lot of water. Here's the formula to calculate your rainfall potential. Gallons of water = catchment area in square feet X rainfall in inches X 0.623.

As long as the roof is somewhat modern, metal or shingles (no old asbestos shingles or nasty chemical sources) that collected water is a goldmine for washing and clothes washing. And it can be purified for drinking as well. The easiest way is to buy a small water purifier for potable water, the type used for camping for example. But there are many water purifiers on the market.

If your home has gutters and a down spout, you are already "collecting" water. All you need to do is divert it to a storage location instead of the water simply running away on the ground unutilized. You can always cut the downspout at a height that allows a trashcan or other buckets to be set under it to collect water.

Roof Water Collection System

Check with your hardware store and purchase a union for your particular downspout profile so that after a return to normal, that cut downspout does not need to be replaced. You merely use the connecting union and join the downspout pieces back together again.

In really desperate times depending on your situation, you could even dig a hole towards the end of your gutter downspout and carefully line it with a plastic sheet to act as an in-ground reservoir.

When it comes down to personal hygiene, a camping shower bag or sponge bath will do the trick. Just as you count on the sun to provide solar electric, that same sun can be used to heat up your water for bathing to something above "frigid." Black camp shower bags for example will heat up nicely through the day and give you a decent shower late afternoon. Use it immediately though since it will cool once the sun stops shining on it.

If you can rig up a cold frame, essentially a box used in cold weather climates for early gardening, setting the shower bag in that will heat the water faster and to a higher temp. The cold frame is just a 4-sided box with an angled lid that has a clear cover which allows the sun to warm the interior of the box. When the box is angled towards the sun, it captures the rays of the sun most efficiently the same as your solar panels do.

If the grid down happens in the winter, use the cold to your advantage. Depending on outdoor temperature, you can freeze or refrigerate your food. When we lived in northern Saskatchewan in the wilderness, we had two chest freezers outdoors.

Outdoor Freezer

When winter showed up, we unplugged them and they were frozen like bricks until spring. We still use ice chests on the porch to occasionally keep food cold. Not only does the ice chest act as a good storage vessel, but with the lid on, it helps protect the food from animal marauders and varmints as well. We've even used clean metal trashcans to store frozen food on the north side of the house during the winter. Geez, I'm really going back in time to the good ol' days.

If it's cold enough overnight to make ice, fill some plastic gallon milk jugs or the equivalent with water and let them freeze. Set them in your refrigerator after they are frozen and they will help keep things cool while at the same time keep the refrigerator compressor from unnecessarily running, thus saving battery power.

Here's another trick we have done with our freezers to conserve energy. Let's say the batteries are fully charged, our charger is regulating meaning it is in float mode, just trickle charging the batteries and in essence we are wasting available power. We take this opportunity to put the freezer on deep freeze, remembering to restore the freezer to its prior setting when the sun goes down. This is another way to conserve battery energy since that super cooled down mass in the refrigerator or freezer will delay the need for the compressor to run for some period of time.

If that is an option on your freezer, turn the thermostat down to a colder temperature so you can utilize the available power to cool that mass of frozen food down further instead of wasting the excess power. Just remember, you must restore the freezer to its prior setting when the sun goes down or you will defeat the purpose of not wasting solar power.

That concept applies if we were running the generator as well. The generator has lots of power. Not only to recharge our batteries with a charger but to run appliances direct off the generator outlets. While the generator is running, you could plug in everything that needs recharging. Then run a cord directly from the generator outlet to the refrigerator or freezer and drop the thermostat down to cool things down. As soon as the generator stops running, just don't forget to plug the refrigerator or freezer back in to the Modular Backyard Power Plant and restore the thermostat to its normal setting.

It might be helpful to make an indelible mark with a magic marker to denote the normal setting so it's easy to return the thermostat to that original position. Once again this is another way to conserve battery energy since that cooled down mass in the refrigerator or freezer will delay the need for the compressor to run for some period of time.

What resources do you have available for food cooking? Unlimited firewood to cook over a campfire? Gas barbecue? Camp stove? Typical briquette barbecue? There are many ways to cook food and I see no need to delve into this too far. But unlike the normal Saturday barbecue, use that available heat the most efficiently. Don't just cook the meal. Use that heat source to cook up food for the next few days. Put the extra in the refrigerator for safe storage for later use.

Another way to use the available heat efficiently is to heat water for dishes or shower using the residual coals to advantage. Set a pot of water on top of the grill lid to heat water. Otherwise, all that residual heat is a wasted heat source.

If you have a Dutch oven, you can cook and even bake cakes and other culinary treats using a campfire.

THE MODULAR BACKYARD POWER PLANT

Campfire Baking and Cooking

A well-fed camper is a happy camper! One can utilize aluminum foil to advantage when cooking over coals too. We've done all of the above and still do some of them from time to time. For example, when I have a barbecue, I always fill the grill with meat. We enjoy a great dinner and the rest is frozen for a quick barbecued meal in the weeks to come. None of these cooking methods work unless you have a store of matches or lighter to ignite a fire. Matches and lighters are cheap. We always inventory some.

I'm just throwing ideas your way to get you thinking about this stuff before you ever need to deal with any of this in a real-life situation. Food goes in via mouth, is processed by the body and...

Regarding toilets… in roughly 43 years, we've never had a flush toilet in our home. And yet, we've survived. We've had outhouses and indoor composting toilets. Those composting toilets divert urine to one place while solids drop into a bucket below. When the bucket gets full, we take it out to a special compost bin where it's left to decompose for at least a year. That composted soil is then used to safely fertilize fruit trees in our orchard.

I cannot know what your individual situation is but I merely point out there are solutions if the flush toilet is unavailable. Mankind was in existence before the discovery of the flush toilet so make do with a bucket or a privacy curtain outdoors surrounding a dug hole. If the public water is off but you have an unlimited source using any of the ideas I've given you above, you can take the toilet tank cover off and fill the reservoir with water and still flush the toilet. Easier still is to pour your water directly into the bowl to flush.

If you have the Modular Backyard Power Plant setup, you have the capability in multiple ways to run lights. As long as the inverter is running, you can plug a normal 120VAC light into the inverter outlet. Keep in mind, if the inverter goes into power save mode, the light will go out which is why you also have the option of running a 24V light direct off the battery.

But a kerosene lamp is also an option.

Kerosene Lantern for Light

Under no circumstances use gasoline. We used kerosene lanterns for many years as our primary light source. Careful attention to wick trimming and cleaning the chimney will provide adequate light. Candles as well will at least provide illumination but I am very leery of open candles. We do not want to compound our troubles by setting the house on fire. Judicious use and 100% attention to any open candles would be the name of the game. Of course, rechargeable flashlights and/or rechargeable batteries for your lights will work wonders as well.

If you decide to inventory kerosene and gasoline, make sure all gasoline is in red containers. Kerosene and diesel should only be stored in yellow containers. That way, there is no chance of a mix up which could be catastrophic. Regardless of whether or not you have your own generator (gas or diesel), your fuel may be a great bartering asset with a neighbor who does have a portable generator. As I've mentioned previously, in a pinch, the charger you have can be used with a generator to fully recharge your batteries. It is just another option for you. If nothing else, that fuel may be a blessing to one of your nearby friends or neighbors.

If power goes down in the heat of summer, there are things to deal with it. The first thing is to realize air conditioning is a relatively new gadget. Humans have lived for millennia without air conditioning and humans in many parts of the world still live without air conditioning. Staying hydrated is vital. Cold drinks will help. Something as simple as a cold washrag draped over the neck will help. Maybe the washrag has a few ice cubes in it to help cool you down. One wants to avoid opening and closing the refrigerator/freezer multiple times so make it count when you go in for something.

In summary, it comes down to making do with what one has to work with. Don't be afraid to think outside the box. Frame and evaluate the problem in your mind and then using all the creativity and materials you either have on hand or can scrounge somewhere, find a short-term solution to get you through. You aren't in this alone. Your friends and neighbors are in the same dilemma. How are they doing things?

The old saying two heads are better than one really comes into play. And guess what. Three heads are better than two. Assuming you have good relations with friends and neighbors, use all their available creativity and knowledge to get you through this. Pool your material things with good neighbors since they might have materials and gadgets that will help everybody equally. Think of each home as being a "store" of goods but when all neighbors pool their "stores" then you have a warehouse of goods.

Your mental attitude will have a huge bearing on how well you endure a longer-term power outage. Mankind survived for millennia without all the modern conveniences. You can too. You will be just fine.

It is my sincere hope that you will find the information and video I've created of value. If nothing else, it will be a resource and knowledge base about solar power as an alternative energy source.

CHAPTER 12
Materials List and Component Sources

The following is a list of the components and parts needed to build the basic *Modular Backyard Power Plant*. This would be the components needed for the *3-Day Blackout Power Plant Module*. I have included in parentheses the quantities needed to buy for the *One-Week Blackout Power Plant Module* and *The Part-Time Power Plant Module*.

It is unlikely one company will have everything you need. My advice is to submit a list to various dealers for pricing. You may find as I did, there can be significant differences on price for the same components from different dealers. Don't be afraid to mention company A has this same product for $50 less, would you match their price. Solar dealers need to make a fair dollar in order to stay in business and provide the components to the public. But being a thrifty consumer is OK too and as individuals, we need to make our hard-earned money go as far as possible. Don't forget to factor in shipping. Some of these companies provide free shipping if the purchase price is over a specified dollar amount.

When inquiring on components whether by phone or email, if you don't get a response or phone call back when you are researching your system parts, my advice is to move on. If you have a battle getting a response when you wish to exchange your hard-earned money for their components, I worry what their service would be like if you actually needed to get help or technical support from them.

Don't panic if the components I specified are no longer available. Things will change over time. I can't know what panels and components will be available in the future, yet, this book/video will be available for years to come. But certain things will never change. The concept of connecting the positive and negative leads to any inverter for example, will be the same. You now have enough knowledge to figure things out. The great thing about solar and electrical systems is one can make changes and adapt if need be.

In October 2022, I selected 200-watt 24-volt panels that were available from multiple sources. But there's no guarantee they will always be available. Let's ponder what can be done if you can't find 200-watt solar panels. We can replace them with something else as long as the total input voltage does not exceed 150V. Ideally the replacements are 24-volt panels but if not ask your local solar dealer's tech support if it's possible to use 12-volt panels in series with your Midnite KID controller.

What panels can you find close to 200 watts? Can you find 100-watt panels for instance? Maybe you can use 8 of them in a series/parallel combination similar to what we are doing with our 200-watt panels. Essentially you would wire two strings of 4 panels with the 2 strings paralleled. Not a great solution because now you are lugging 8 panels outdoors when the grid goes down but this is a possible solution.

Can you find 130-watt panels? You might be able to wire 2 strings of 3 panels paralleled. I say might because I can't know what you will find and what the voltages will be for the panels you find. That is a technical question best put to tech support.

Here is the information you need to tell them. You have one Midnite KID solar controller. You want to buy the brand XYZ solar panels they have on sale. Ask them how many can you buy and how do they suggest hooking them up to fully utilize the Midnite KID capabilities. Keep in mind, I chose the 200-watt panel because it does fully utilize the capabilities of this KID controller. Can you hook up a couple 300-watt panels? Perhaps but it depends on whether the output voltage is below the 150V maximum. A 300-watt panel is getting big to lug around which is another issue. But more importantly, the KID controller can handle 800 watts and you only supply 600 watts so you wouldn't be fully utilizing this controller. That will significantly impact charging your battery back up to full.

For those interested in the more technical side of things. You can visit the midnite website: https://www.midnitesolar.com/sizingTool_kid/index.php. You can insert the specs of the panel you are thinking of purchasing and it will tell you if it will work. As you can see, all I can do is throw ideas your way since I cannot know what panels and specifications you will find down the road. But the point is, you will have options. This project will never become obsolete as long as one adapts the components available in the future. I've put together a combination of components I feel will serve the most people the best for the intended purpose of getting through a grid down situation of unknown duration.

As well, this applies to the batteries and other components. Things change. Electronics and the technologies improve. You are not locked into any particular brand or battery type. If another battery company has a 24 VDC lithium battery on sale, there is no reason you can't buy it as long as it is 24V, can be hooked up in parallel and provides the amp hour storage you require. You may have to tweak the charge controller settings for your particular battery but that's very easy to do via the control buttons and information on the display.

I'm sure there are many good companies out there but the following companies would be a good place to start when shopping for components. I have no affiliation with the following companies. All I can tell you is they were responsive and I was able to go back and forth with the sales reps to get questions answered.

If you are in Canada, I can vouch for two companies since I have purchased components from them for our Solar power system we use here in Nova Scotia. Both companies and the guys have been ultra responsive, ultra helpful and they back up the products they sell. They sell everything needed for solar electric power systems.

Rob Beckers at Solacity (https://www.solacity.com/) is owed a special acknowledgment and thank you. I've never met Rob but we've exchanged a lot of emails over the last 6 years and I feel like I've known him far longer than those 6 years. Rob kindly read the electrical chapter in our previous book The Self-Sufficient Backyard: For the Independent Homesteader and made helpful suggestions to improve it.

As well, **John Vanden Broek at Hub Power** (https://hub-power.ca/) furnished some input on batteries for this book. My thanks to John. Both Rob at Solacity and John at Hub Power are as sharp and as knowledgeable as they come.

Here are the companies in the United States I dealt with. Each of the named people were responsive, knowledgeable and helpful. The companies are in no particular order:

Brian Betz @ Backwoods Solar
(https://backwoodssolar.com/)
Complete inventory of solar components.

Mike Ridden @ The Solar Store
(http://www.thesolarstore.com/)
Complete inventory of solar components.

Jason Weber @ Northern Arizona Wind and Sun
(https://www.solar-electric.com/)
Complete inventory of solar components.

Shaylene Marino @ altestore
(https://www.altestore.com/)
Complete inventory of solar components. As well, this company is the source for the 200-watt solar panels for the power plant.

Brooke Flecknell @ unplugged Power
(https://unpluggedpowersystems.ca/)
is a Canadian source for the Power/USB module. There are other sources as well in the US. Companies that supply RV and Marine equipment will have a module similar.
https://www.waytekwire.com/item/11014/11014-Dual-2-1A-USB-Charger-/

Amber Sparrow @ watt-a-light
(https://wattalight.com/)
is the source for anything related to LED lighting. Both AC and DC but they carry a wide array of DC light bulbs in different bases and voltages.

If you ever contact any of the listed sources, tell them Ron sent you. When they say Ron who, you say, "you know, the swell guy in Nova Scotia who wrote that off grid book."

Maybe they'll remember me, maybe they won't. But really... how could they forget me?

THE MODULAR BACKYARD POWER PLANT

On behalf of Johanna and myself, we wish you all the best with your new power plant. Although we hope it will rarely if ever be used, it's an insurance investment and mental reassurance that should the grid go down for whatever reason, you will have the resources to get through the ordeal with the least amount of chaos and disruption.

Qty	Descripion	Item Number	Notes	
1 (2,4)	Renogy 24V 50Ah Lithium Iron Phosphate Battery *LiFePO4*	SKU: RBT2450LFP-US	$410 ea	820
0 (2,6)	Battery Interconnect Cable – 10⅜ – #1 AWG w/ ⅜" lugs			
(1)	24v 20 Amp Dakota Lithium LiFePO4 Battery Charger		https://dakotalithium.com/product/ultra-fast-24v-20-amp-dakota-lithium-lifepo4-battery-charger/	125
1	COTEK SD2500-124 2500 Watt 24 Volt Pure Sine Wave GFCI Inverter	SKU SD2500-124 GFCI		810
1 (2,4)	Rich Solar MEGA 200 Watt Monocrystalline Solar Panel 24V	SKU RS-M200D or ALTS200-24P $219	https://www.altestore.com/store/solar-panels/alte-solar-panels-p40768/	440
1	MidNite Solar The Kid MPPT Solar Charge Controller in Black	SKU MNKID-B		380
1	Mini-DC Disconnect Power Center 175A MNDC175 (MNDC)	Model: 053-00092		280
1	MNEPV Midnite Solar MNEPV 1 to 63 Amp 150 VDC Breakers for Solar Panel Arrays	MNEPV-20 DIN Rail Mount		25
1	MC4 Solarline 2 PV Array Output Cable 70' #10 AWG	Model: 052-09417		

Qty	Descripion	Item Number	Notes
1	#1 AWG 5' red flexible wire battery cable w/ ⅜" lugs both ends	Sometimes come as a pair	
1	#1 AWG 5' black flexible wire battery cable w/ ⅜" lugs both ends	Sometimes come as a pair	
1	Inverter cable 1 AWG, 3 ft flexible wire, red with 3/8" lugs	Sometimes come as a pair	
1	Inverter cable 1 AWG, 3 ft flexible wire, black with ⅜" lugs	Sometimes come as a pair	
1	MC4, SOLARLINE2, LOCKING CONN, OPEN-END SPANNER WRENCH SET (2)	Model: 094-00112	
1	12 inch #10 AWG jumper with red insulation; 1 end w/⅜" crimped ring lug.		
1	24 inch #10 AWG jumper with black insulation		
1	22 inch #10 AWG jumper with red insulation		
1	22 inch #10 AWG jumper with white or red insulation		
0 (0,1)	Multi-Contact 32.0018 PV-AZB4, MC4 Y-connector, (1) male / (2) female, 1500V, 50A		"Y" Connectors only needed for The Part-Time Power Plant Module
0 (0,1)	Multi-Contact 32.0019 PV-AZS4, MC4 Y-connector, (2) male / (1) female, 1500V, 50A		"Y" Connectors only needed for The Part-Time Power Plant Module
2	Large strain reliefs for DC Mini Enclosure and 2 small strain reliefs for the Charge Controller.		

note 1: I specified 70 feet of solar wire. If longer length needed, order appropriate length.